NeuroCognition Revealed

The Keys to Understanding
the Mind

Reynard Windrow

Chapter 1: Foundations of Neural Architecture

Basic Building Blocks Neurons and Synapses

The human brain, a marvel of biological engineering, is composed of an intricate network of cells that form the foundation of all neural activity. At the core of this complexity are neurons and synapses, the basic building blocks of the nervous system. Neurons, often referred to as nerve cells, are the primary units responsible for transmitting and processing information throughout the body. They operate in a delicate balance of electrical and chemical signals, forming the essential framework for thought, sensation, and action.

Each neuron is a specialized cell that consists of three primary components: the soma, or cell body; dendrites, which extend like branches from the soma; and the axon, a long, slender projection that carries signals away from the cell body. The soma houses the nucleus, which contains the genetic material necessary for the cell's functioning and maintenance. Surrounding the nucleus is the cytoplasm, where vital organelles such as the mitochondria and endoplasmic reticulum support the energy demands and protein synthesis required for neural activity.

Dendrites play a crucial role in receiving incoming signals from other neurons. Their branching structure increases the surface area of the neuron, allowing it to form connections with multiple other cells. These connections, known as synapses, are the points where communication occurs between neurons. Synapses can form on the dendrites, soma, or even along the axon itself. The axon, in contrast, serves as the primary conduit for transmitting electrical impulses, known as action potentials, to other neurons or target tissues. At the end of the axon, terminal branches culminate in synaptic boutons, which house the machinery necessary for chemical communication.

The transmission of information within a neuron is an electrochemical process. It begins with the generation of an action potential, a rapid change in electrical charge across the neuronal membrane. This occurs when the neuron is sufficiently stimulated, causing voltage-gated ion channels to open and allowing the influx of positively charged sodium ions. The resulting depolarization propagates along the length of the axon, ultimately reaching the synaptic terminal. Here, the electrical signal is converted into a chemical signal through the release of neurotransmitters.

Synapses, the junctions where neurons communicate, can be broadly classified into two types: chemical and electrical. Chemical synapses are the most common and involve the release of neurotransmitters into the synaptic cleft, a small gap between the presynaptic and postsynaptic neurons. These chemical messengers bind to specific receptors on the postsynaptic membrane, triggering a response that may either excite or inhibit the postsynaptic neuron. Electrical synapses, on the other hand, involve direct transmission of electrical currents through gap junctions, allowing for faster and more synchronized communication. Though less common in humans, electrical synapses are prevalent in certain brain regions and play a role in processes requiring rapid coordination.

The diversity of neurotransmitters and their receptors adds another layer of complexity to neural communication. Each neurotransmitter has specific effects depending on the type of receptor it binds to. For instance, glutamate is the primary excitatory neurotransmitter in the brain, while gamma-aminobutyric acid (GABA) serves as the main inhibitory counterpart. Dopamine, serotonin, acetylcholine, and norepinephrine are examples of modulatory neurotransmitters that influence mood, attention, and other higher-order functions. The balance of excitatory and inhibitory signals at the synapse is critical for maintaining neural stability and preventing disorders such as epilepsy.

Neurons do not function in isolation; they are organized into networks that enable complex processing and integration of information. These networks are shaped by the connectivity of synapses, which can vary in strength and adaptability. Synaptic plasticity, the ability of synapses to strengthen or weaken over time, is a fundamental mechanism underlying learning and memory. Long-term potentiation (LTP) and long-term depression (LTD) are two well-studied forms of synaptic plasticity that illustrate how repeated activity can lead to lasting changes in synaptic strength. These modifications are thought to enhance the efficiency of neural circuits, allowing them to store and retrieve information more effectively.

Another critical aspect of neural function is the support provided by glial cells. Though often overshadowed by neurons, glial cells are indispensable for maintaining the health and efficiency of the nervous system. Astrocytes, a type of glial cell, regulate the extracellular environment, provide metabolic support to neurons, and assist in the formation and maintenance of synapses. Oligodendrocytes and Schwann cells, in contrast, produce myelin, a fatty substance that insulates axons and accelerates the conduction of action potentials. Microglia, the immune cells of the brain, play a role in clearing debris and responding to injury or infection.

The dynamic interplay between neurons, synapses, and glial cells forms the basis of all neural activity. This intricate system is further modulated by factors such as gene expression, environmental influences, and developmental processes. From the microstructure of individual synapses to the macrostructure of neural networks, the organization of the nervous system reflects the remarkable adaptability and resilience of the human brain. Understanding the fundamental building blocks of neurons and synapses offers a window into the mechanisms that drive cognition, emotion, and behavior, laying the groundwork for advances in neuroscience and medicine.

Neural Networks and Circuit Formation

Neural networks are the foundation of the brain's ability to process information, enabling everything from basic reflexes to complex reasoning. These networks consist of interconnected neurons that communicate through synapses, forming circuits that dictate how information flows and is integrated within the brain. The formation of these circuits is a highly ordered yet dynamic process, influenced by genetic programming, environmental input, and ongoing neural activity. Each circuit is meticulously designed to serve specific functions, whether it is regulating involuntary bodily processes, interpreting sensory data, or executing voluntary movements.

The architecture of neural networks begins with the establishment of connectivity during early development. Neurons are generated through neurogenesis, a process that occurs primarily during the embryonic and early postnatal stages. Once formed, neurons migrate to their designated locations within the brain, guided by chemical gradients and structural cues. This migration ensures that neurons are positioned in the appropriate regions to later form functional circuits. As neurons settle into their positions, they extend axons and dendrites, the structures responsible for sending and receiving signals. Axons grow toward their targets by following guidance cues in their environment, while dendrites branch out to form numerous connections.

The formation of synapses—the junctions where neurons communicate—is a critical step in building neural circuits. Synaptogenesis, the process of synapse formation, involves the precise alignment of presynaptic and postsynaptic structures. Specialized proteins facilitate this alignment, ensuring that neurotransmitters released from the presynaptic neuron can effectively bind to receptors on the postsynaptic membrane. Initially, an overabundance of synapses is formed, creating a surplus of connections. This overproduction is followed by synaptic pruning, a process in which unused or inefficient synapses are eliminated. Pruning refines neural circuits, enhancing their efficiency and specificity by strengthening frequently used synapses while discarding redundant ones.

Activity-dependent processes play a significant role in shaping neural networks during development and throughout life. Neural activity, driven by sensory experiences and interactions with the environment, reinforces the connections that are most relevant to an individual's experiences. For example, during critical periods of development, sensory inputs such as visual or auditory stimuli are essential for the proper wiring of the corresponding brain regions. If these inputs are restricted or absent, the affected neural circuits may fail to develop fully, leading to long-lasting deficits. This principle, known as "use it or lose it," underscores the importance of environmental interactions in shaping the brain's architecture.

Distinct types of neural circuits are specialized for different functions, and their organization reflects the hierarchical nature of brain processing. Some circuits operate locally, confining their activity to specific brain regions, while others span multiple areas to integrate information across the brain. For instance, sensory circuits process data from the external environment, transforming raw sensory input into meaningful perceptions. Motor circuits, on the other hand, translate decisions made in higher-order areas into physical actions. Between these extremes, associative circuits link sensory and motor information, enabling complex behaviors such as decision-making and problem-solving.

The brain's ability to form circuits that adapt to new information is a testament to its plasticity. Neural networks are not static; they continuously reorganize in response to learning and experience. Long-term potentiation (LTP) and long-term depression (LTD) are two well-documented mechanisms that underpin this adaptability. LTP strengthens synaptic connections through repeated use, making it easier for signals to pass between neurons. LTD, conversely, weakens connections that are rarely used, freeing up resources for more relevant pathways. Together, these processes ensure that the brain remains flexible, capable of acquiring new skills and adapting to changing environments.

In addition to their adaptability, neural networks exhibit a remarkable degree of specialization. Different brain regions are dedicated to specific functions, and the circuits within these areas are tailored to their roles. The visual cortex, for example, contains circuits that process various aspects of visual information, such as color, motion, and depth. Similarly, the motor cortex houses circuits that control precise movements, while the prefrontal cortex is involved in planning and decision-making. Despite this specialization, neural circuits do not operate in isolation. Instead, they are interconnected, forming larger networks that enable the brain to coordinate complex behaviors and cognitive processes.

The development and maintenance of neural circuits are supported by a variety of cellular and molecular mechanisms. Glial cells, often overshadowed by neurons, play a critical role in this process. Astrocytes, a type of glial cell, regulate the formation and elimination of synapses, ensuring the proper balance of connectivity within circuits. Oligodendrocytes produce myelin, which insulates axons and speeds up signal transmission, enhancing the efficiency of neural communication. Microglia, the brain's immune cells, contribute to synaptic pruning by engulfing and removing inactive synapses. These supporting cells work in concert with neurons to establish and sustain functional networks.

Disruptions in neural circuit formation can have profound consequences for brain function. Genetic mutations, environmental factors, and injuries can interfere with the processes of synaptogenesis, pruning, or plasticity, leading to a range of neurological and psychiatric disorders. For example, imbalances in synaptic pruning have been implicated in conditions such as autism spectrum disorder and schizophrenia. Understanding the mechanisms that govern circuit formation provides valuable insights into these disorders and opens the door to potential therapeutic interventions.

The interplay between genetic programming and environmental influence is at the heart of neural network formation. While genes provide the blueprint for the brain's initial wiring, experience and activity fine-tune these connections, ensuring that the brain remains responsive to the demands of its environment. This dynamic process, which continues throughout life, highlights the adaptability and resilience of the human brain. By unraveling the complexities of neural networks and circuit formation, researchers gain a deeper understanding of how the brain functions, paving the way for advancements in neuroscience and medicine.

Brain Regions and Their Specializations

The human brain is an intricate organ composed of distinct regions, each specialized for specific functions, yet seamlessly interconnected to facilitate the complexity of human thought, emotion, and behavior. Its organization reflects billions of years of evolution, with each region contributing unique capabilities to the overall system. Understanding these regions and their specializations provides a glimpse into the mechanisms that govern everything from sensory perception to abstract reasoning.

The cerebral cortex, the brain's outermost layer, is perhaps its most iconic structure. This wrinkled, folded expanse of gray matter is divided into two hemispheres, each of which controls the opposite side of the body. Within the cortex, four primary lobes—frontal, parietal, occipital, and temporal—serve as the command centers for various functions. The frontal lobe, located at the front of the brain, is responsible for higher cognitive processes like decision-making, problem-solving, and planning. It is also home to the motor cortex, which directs voluntary movements, and Broca's area, a region crucial for speech production. Damage to the frontal lobe often results in profound changes in personality, behavior, and the ability to execute complex tasks.

The parietal lobe, situated behind the frontal lobe, integrates sensory information from across the body. Its primary feature, the somatosensory cortex, processes touch, temperature, pain, and proprioception—the sense of body position. The parietal lobe also plays a role in spatial awareness and navigation, allowing an individual to orient themselves within their environment. Adjacent to the parietal lobe, the occipital lobe at the back of the brain is dedicated to vision. It houses the primary visual cortex, where raw visual data from the eyes is processed and interpreted. Damage to this region can result in partial or complete blindness, even if the eyes themselves remain functional.

The temporal lobe, located beneath the frontal and parietal lobes, is essential for auditory processing and memory. Within this region lies the primary auditory cortex, which decodes sound signals from the ears, as well as Wernicke's area, a critical structure for language comprehension. The temporal lobe is also deeply connected to the hippocampus, a structure vital for the formation and retrieval of memories. This intricate relationship between memory and sensory input highlights the interconnected nature of brain regions, as stimuli from the environment are stored and recalled to inform future actions and decisions.

Beneath the cerebral cortex lies the subcortical brain, a collection of structures that provide essential support for both basic survival and higher-order functions. The thalamus, often described as the brain's relay station, channels sensory information to the appropriate cortical areas for processing. Its central location and connections to nearly every part of the brain make it indispensable for integrating sensory and motor signals. Nearby, the hypothalamus regulates critical autonomic functions such as hunger, thirst, sleep, and body temperature. It also governs the release of hormones via its control over the pituitary gland, linking the nervous system to the endocrine system.

The limbic system, nestled deep within the brain, is central to emotion, motivation, and memory. Key components of this system include the amygdala and hippocampus. The amygdala is involved in processing emotions such as fear, anger, and pleasure, and it plays a role in forming emotional memories. The hippocampus, meanwhile, is essential for converting short-term memories into long-term ones and navigating spatial environments. Together, these structures ensure that experiences are not only remembered but also imbued with emotional significance.

The brainstem, located at the base of the brain, is the most primitive region, responsible for fundamental life-support functions. Divided into the midbrain, pons, and medulla oblongata, the brainstem regulates breathing, heart rate, and blood pressure. It also acts as a conduit for information traveling between the body and the higher brain regions. The cerebellum, located behind the brainstem, is often overshadowed by the larger cerebral cortex but is no less vital. It coordinates movement, balance, and posture, refining motor commands to ensure precision and efficiency. Recent research has also revealed its involvement in certain cognitive and emotional processes, further underscoring its importance.

The basal ganglia, a group of nuclei situated deep within the brain, are pivotal for motor control and learning. These structures facilitate the initiation and regulation of voluntary movements, ensuring smooth and purposeful actions. Dysfunction within the basal ganglia is associated with movement disorders such as Parkinson's disease and Huntington's disease, which highlight its critical role in motor function. Beyond movement, the basal ganglia contribute to habit formation and reward processing, linking them to motivation and behavior.

Despite the delineation of these specialized regions, the brain functions as an integrated whole, with constant communication between its parts. Neural pathways connect distant regions, enabling the seamless exchange of information. For instance, the corpus callosum, a thick bundle of nerve fibers, bridges the two hemispheres, allowing them to work in unison. Similarly, the connections between the limbic system and the prefrontal cortex facilitate the regulation of emotions through reasoned thought.

The brain's remarkable adaptability, or plasticity, further blurs the lines between specialization and integration. When one region is damaged, other areas can often compensate by reorganizing and taking over its functions. This ability to adapt underscores the brain's resilience and highlights its capacity for learning and recovery.

Each region of the brain contributes to the intricate symphony of human experience, from the automatic regulation of vital functions to the conscious deliberation of abstract ideas. The specialization of these regions ensures efficiency, while their interconnectedness guarantees fluidity and coherence in thought and action. This dynamic interplay of structure and function reveals the extraordinary complexity of the brain, an organ that continues to fascinate and inspire exploration.

Neuroplasticity The Adaptive Brain

The human brain is a dynamic and adaptive organ, capable of remarkable transformations in response to experience, injury, learning, and environmental changes. This adaptability, known as neuroplasticity, is the brain's ability to reorganize itself by forming new neural connections, strengthening existing ones, or even rerouting functions to entirely different regions. Far from being a static structure, the brain remains in a perpetual state of flux, constantly reshaping itself to optimize performance and meet the demands of life.

Neuroplasticity operates through two primary mechanisms: structural and functional changes. Structural plasticity refers to physical changes in the brain's architecture, such as the growth of new dendrites, the formation of additional synapses, or even the birth of new neurons—a process called neurogenesis. Functional plasticity, on the other hand, involves the brain's ability to shift functions from one area to another, often observed when an injury or disease impairs a particular region. These two mechanisms work in tandem, ensuring that the brain remains resilient and adaptable in the face of challenges.

One of the most striking examples of neuroplasticity can be seen in the way the brain responds to learning and memory formation. Each time a new skill is learned or a piece of information is acquired, the brain undergoes subtle yet measurable changes. Synapses, the points of communication between neurons, are strengthened through repeated use—a process known as long-term potentiation. This strengthening enhances the efficiency of neural circuits, making it easier for the brain to recall or execute what has been learned. Conversely, synapses that are rarely used may weaken over time, a phenomenon called long-term depression. This selective pruning of unused connections helps to optimize the brain's resources, ensuring that energy and space are allocated to the most relevant functions.

Neuroplasticity is particularly pronounced during certain periods of life, such as early childhood. The developing brain is a hotbed of plasticity, forming millions of new connections each day as it absorbs vast amounts of sensory, cognitive, and emotional information. This heightened plasticity allows children to acquire language, motor skills, and social behaviors rapidly and with remarkable ease. However, the brain's capacity for change does not vanish with age. While it may become less malleable over time, the adult brain retains significant plasticity, particularly in response to focused effort and practice. This is why adults can still learn new languages, develop new habits, or recover from injuries through rehabilitation.

The concept of neuroplasticity extends beyond learning and development; it also plays a critical role in recovery from brain injuries. When a specific region of the brain is damaged—whether from a stroke, traumatic injury, or neurological disease—neighboring areas or even distant regions can sometimes take over its functions. This phenomenon, known as cortical reorganization, is a testament to the brain's resilience. For example, individuals who lose function in one hemisphere of the brain may, over time, regain certain abilities as the opposite hemisphere compensates. Rehabilitation therapies, such as physical or occupational therapy, often leverage neuroplasticity by encouraging repetitive practice and targeted exercises, which reinforce alternative neural pathways and promote recovery.

Plasticity is not limited to the reorganization of regions; it also extends to sensory modalities. An extraordinary example is observed in individuals who are blind. Without visual input, their brains often repurpose the visual cortex for other functions, such as processing auditory or tactile information. This adaptation enhances their ability to read Braille or navigate using sound, demonstrating how the brain reallocates resources to maximize capabilities in the absence of a particular sense. Similarly, musicians who dedicate years to their craft often exhibit structural changes in brain areas associated with fine motor control and auditory processing, underscoring how focused practice can reshape neural circuits.

While neuroplasticity is a powerful and beneficial phenomenon, it is not inherently positive in all cases. The same mechanisms that enable the brain to adapt and learn can also reinforce maladaptive patterns. For instance, chronic stress or trauma can lead to the strengthening of neural pathways associated with fear and anxiety, making these responses more easily triggered in the future. Similarly, addictive behaviors can hijack the brain's reward system, reinforcing circuits that perpetuate dependency. These examples highlight the dual nature of neuroplasticity: its capacity to drive both growth and dysfunction, depending on the context.

The molecular and cellular basis of neuroplasticity is equally fascinating. Changes in synaptic strength are mediated by alterations in the release of neurotransmitters, the sensitivity of receptors, and the structural integrity of synapses. Additionally, neurotrophic factors, such as brain-derived neurotrophic factor (BDNF), play a crucial role in promoting the growth and survival of neurons, as well as supporting the formation and maintenance of synapses. Physical exercise, mental stimulation, and a healthy diet have all been shown to increase levels of BDNF, suggesting that lifestyle choices can significantly influence the brain's plastic potential.

The implications of neuroplasticity extend far beyond the individual. It has reshaped the way scientists, educators, and clinicians approach brain health, learning, and recovery. For educators, understanding that the brain can rewire itself through practice and repetition has led to the development of teaching methods that emphasize active engagement and adaptability. For clinicians, the recognition of neuroplasticity has revolutionized approaches to treating neurological disorders, from stroke rehabilitation to therapies for conditions like depression or post-traumatic stress disorder. Even in aging populations, interventions such as cognitive training and physical activity are being used to harness neuroplasticity, mitigating the effects of cognitive decline and enhancing quality of life.

The brain's capacity for change is a cornerstone of human resilience, allowing individuals to adapt to new environments, overcome challenges, and continually grow throughout their lives. It embodies the essence of human potential, illustrating how even the most ingrained patterns of thought or behavior can be altered with effort and intention. By understanding and leveraging the principles of neuroplasticity, individuals have the opportunity to not only recover lost functions but also unlock new levels of capability and creativity, demonstrating the extraordinary adaptability of the human brain.

Evolution of Neural Systems

Neural systems, the networks of cells that process and transmit information, have undergone billions of years of evolution, shaping the remarkable complexity of the human brain we know today. This evolutionary journey reflects the gradual adaptations that allowed organisms to respond more effectively to their environments, survive changing conditions, and develop increasingly sophisticated behaviors. From the simplest nerve nets in ancient sea creatures to the intricate neural circuits that govern human cognition, the progression of neural systems is a testament to the power of adaptation and innovation in the natural world.

The earliest forms of nervous systems emerged in simple, multicellular organisms over 600 million years ago. These primitive systems were composed of basic nerve nets, such as those found in cnidarians like jellyfish. Nerve nets lack centralized structures; instead, they consist of a diffuse network of neurons that communicate directly with one another. This design allowed basic motor responses, such as contracting muscles to propel through water or reacting to touch. Although rudimentary, these nerve nets represented a significant evolutionary step, enabling organisms to coordinate movement and respond to external stimuli in a way that single-celled organisms could not.

As evolution progressed, more complex nervous systems began to appear. Bilateral symmetry in animals, a major evolutionary milestone, brought about the centralization of neural structures. This symmetry allowed for the development of nerve cords, which ran along the length of the body, and the concentration of neurons in specific regions called ganglia. Flatworms, for example, exhibit this early form of centralization, with a simple brain-like structure and paired nerve cords. These adaptations enabled more coordinated movements and the processing of sensory information, offering a survival advantage in more dynamic environments.

The next major leap in neural evolution came with the emergence of segmented nerve cords in annelids and arthropods. These segmented systems allowed for greater specialization and efficiency in processing information. Each segment had its own ganglia, which could independently control local movements, while the central brain handled more complex tasks. This division of labor marked the beginning of hierarchical organization within nervous systems, a feature that persists in more advanced organisms. Arthropods, such as insects, further refined this structure, evolving sophisticated sensory organs like compound eyes and antennae, which provided detailed input to their central nervous systems.

The evolution of vertebrates brought about a dramatic expansion and specialization of neural systems. The dorsal nerve cord, a defining characteristic of the vertebrate lineage, became enclosed within a protective spinal column. At the anterior end of this cord, the brain began to develop into distinct regions, each dedicated to specific functions. Early vertebrates, such as jawless fish, exhibited basic brain structures divided into the forebrain, midbrain, and hindbrain. These regions laid the groundwork for higher-order processing, motor coordination, and autonomic control.

As vertebrates diversified, their nervous systems adapted to the demands of their environments. Fish, for instance, developed an enlarged hindbrain to improve balance and coordination in aquatic settings. Amphibians, transitioning to life on land, refined sensory systems to navigate both terrestrial and aquatic habitats. Reptiles, with their increased reliance on vision and complex motor patterns, exhibited further brain specialization, particularly in the forebrain. This trend of forebrain development continued in birds and mammals, culminating in the highly intricate cerebrum of primates.

The cerebral cortex, a hallmark of mammalian evolution, represents one of the most significant advancements in neural systems. This outer layer of the brain is responsible for higher cognitive functions such as reasoning, memory, and voluntary movement. Early mammals relied on their expanded cortex for enhanced sensory processing, particularly in olfaction, which was critical for survival in nocturnal environments. Over time, the cortex evolved to support more complex behaviors, including social interactions and problem-solving.

Primates, including humans, experienced unparalleled growth in the neocortex, the part of the brain associated with advanced cognitive abilities. This expansion allowed for the development of abstract thinking, language, and self-awareness. The folding of the cortex into gyri and sulci increased its surface area, accommodating a greater number of neurons without requiring a larger skull. This adaptation facilitated the emergence of behaviors that define humanity, such as art, science, and culture.

The evolution of neural systems did not occur in isolation but was shaped by interactions with other biological systems and environmental pressures. For example, the development of sensory organs, such as eyes and ears, drove the refinement of neural pathways dedicated to visual and auditory processing. Similarly, the evolution of limbs and fine motor skills necessitated more sophisticated motor control systems. Social structures and communication further influenced the complexity of neural circuits, particularly in species that relied on group cooperation for survival.

One of the most remarkable aspects of neural evolution is the retention of ancient structures alongside the development of new ones. The human brain, for example, still contains the brainstem and limbic system, which control basic survival functions and emotions, inherited from our distant ancestors. These older structures coexist with the neocortex, forming an integrated system that balances primal instincts with advanced reasoning.

Despite its many adaptations, the evolution of neural systems was not without trade-offs. The increased complexity and energy demands of larger brains required significant metabolic resources, which in turn influenced dietary and social behaviors. Additionally, the reliance on advanced neural systems made organisms more vulnerable to disruptions, such as injury or disease. However, the advantages conferred by these systems far outweighed their costs, driving the continued refinement of neural structures over millions of years.

The journey of neural evolution illustrates the profound interplay between biology and environment.

From the simplest nerve nets to the intricate circuitry of the human brain, each step in this progression reflects the challenges and opportunities faced by organisms throughout history. This evolutionary process not only shaped the diversity of life on Earth but also laid the foundation for the extraordinary capabilities of human thought, creativity, and adaptation. By tracing the origins and development of neural systems, we gain a deeper appreciation for the intricate mechanisms that underpin all aspects of life.

Chapter 2: Cognitive Processes Unveiled

Processing Speed and Information Flow

The human brain processes vast amounts of information every second, orchestrating a symphony of electrical and chemical signals that govern our thoughts, actions, and perceptions. At the heart of this extraordinary capability lies the brain's processing speed and the efficiency of information flow within its intricate neural networks. These functions are critical not only for basic survival but also for the higher-order cognitive abilities that distinguish humans from other species. The speed and accuracy of these processes determine how we interact with the world, solve problems, and adapt to ever-changing environments.

Processing speed refers to the rate at which the brain can interpret and respond to incoming stimuli. It is influenced by the electrical properties of neurons, the quality of synaptic communication, and the integrity of the pathways that connect different brain regions. Information flow, on the other hand, describes the movement of data through these neural pathways, enabling the coordination of sensory input, memory, decision-making, and motor output. Together, these mechanisms form the backbone of cognitive function, allowing the brain to operate with remarkable precision and agility.

The foundation of processing speed lies in the action potential, the electrical signal that travels along neurons. This signal is generated when a neuron is stimulated, causing a rapid influx of positively charged ions into its membrane. The resulting depolarization triggers a chain reaction, with the signal propagating down the axon toward the synaptic terminal. The speed of this transmission depends on several factors, including the diameter of the axon and the presence of myelin, a fatty substance that insulates the axon and accelerates signal conduction. Larger axons and those with thicker myelin sheaths transmit signals more quickly, ensuring rapid communication between neurons.

The efficiency of information flow is further enhanced by the organization of neural pathways. In the brain, neurons are not randomly connected; instead, they are arranged into specialized networks optimized for specific functions. These networks are connected by white matter tracts, bundles of myelinated axons that act as highways for signal transmission. The quality of these tracts plays a crucial role in determining how quickly and accurately information can travel between different regions of the brain. For example, the corpus callosum, the largest white matter structure, facilitates communication between the left and right hemispheres, enabling the integration of sensory and motor information.

While the speed of individual action potentials is important, the brain's true power lies in its ability to process information in parallel. Unlike a computer that performs one calculation at a time, the brain can handle multiple tasks simultaneously by distributing them across various neural circuits. This parallel processing allows for the rapid integration of sensory inputs, such as sight, sound, and touch, into a cohesive perception of the environment. It also enables the brain to make split-second decisions, coordinating complex motor actions while simultaneously evaluating potential consequences.

The brain's capacity for rapid processing and information flow is not uniform across all regions. Some areas, such as the primary sensory cortices, are designed for high-speed input processing. The visual cortex, for instance, can interpret incoming data from the eyes in milliseconds, allowing us to react almost instantaneously to changes in our surroundings. Other regions, like the prefrontal cortex, are involved in more deliberate, slower processes, such as reasoning, planning, and decision-making. This balance between speed and deliberation ensures that the brain can respond appropriately to both immediate and long-term challenges.

Processing speed and information flow are not fixed attributes; they can be influenced by a variety of factors, including age, experience, and health. During childhood and adolescence, the brain undergoes significant development, with the myelination of axons and the strengthening of synaptic connections enhancing processing speed. As individuals gain experience and practice specific skills, the efficiency of neural pathways improves, allowing for faster and more accurate performance. For example, musicians and athletes often demonstrate superior processing speeds in tasks related to their expertise, a result of years of dedicated practice and neural refinement.

However, the brain's processing capabilities can decline with age and in the presence of certain neurological conditions. The degradation of myelin, the loss of synaptic connections, and reduced blood flow to the brain are common factors that contribute to slower processing speed in older adults. Disorders such as multiple sclerosis, Alzheimer's disease, and traumatic brain injuries can further impair information flow, leading to cognitive deficits and functional limitations. Despite these challenges, the brain retains a remarkable degree of plasticity, and interventions such as cognitive training, physical exercise, and a healthy diet can help maintain or even improve processing speed and information flow.

The importance of efficient processing and information flow extends beyond individual neurons and pathways. At the level of large-scale brain networks, the coordination of activity between distant regions is essential for complex cognitive functions. The default mode network, for instance, is active during rest and introspection, while the central executive network governs attention and goal-directed behavior. The dynamic interplay between these networks relies on the rapid exchange of information, highlighting the importance of global connectivity for overall brain function.

Disruptions in processing speed and information flow can have profound effects on cognition and behavior. For example, in conditions such as attention deficit hyperactivity disorder (ADHD) and autism spectrum disorder (ASD), atypical connectivity patterns and slowed information processing can contribute to difficulties in attention, communication, and social interactions. Similarly, in individuals with stroke or traumatic brain injury, damage to white matter tracts can impair the flow of information between key brain regions, leading to deficits in motor control, language, and memory.

Despite the challenges posed by these disruptions, the brain's inherent adaptability offers hope for recovery and improvement. Through targeted interventions, such as rehabilitation therapies and neurofeedback, individuals can enhance their processing speed and restore efficient information flow. These approaches leverage the brain's capacity for plasticity, encouraging the formation of new connections and the strengthening of existing ones.

The interplay between processing speed and information flow underscores the brain's extraordinary complexity and efficiency. By understanding the mechanisms that drive these functions, researchers can develop strategies to optimize cognitive performance, enhance learning, and mitigate the effects of aging and disease. The rapid and seamless coordination of neural activity is what allows us to navigate the world, solve problems, and adapt to the ever-changing demands of life, making it one of the most remarkable features of the human brain.

Memory Formation and Retrieval

Memory is the intricate process through which the brain encodes, stores, and retrieves information, enabling us to learn, adapt, and navigate the world. It serves not only as a repository of past experiences but also as a framework for anticipating the future and making informed decisions. From the vivid recollection of a childhood event to the unconscious recognition of a familiar face, memory operates through a seamless interplay of complex neural mechanisms.

The process of memory formation begins with encoding, the initial stage where sensory input is transformed into a construct the brain can store. This transformation involves attention, as the brain filters vast amounts of sensory information to focus on stimuli deemed significant. For example, when meeting someone new, you may focus on their name, appearance, and voice. These details are processed by the sensory regions of the brain and then relayed to the hippocampus, a structure essential for consolidating memories. The hippocampus acts as a hub, integrating information from various sensory modalities and preparing it for long-term storage.

Not all memories are created equal. Some are fleeting and quickly forgotten, while others persist for years. This distinction arises from differences in how memories are consolidated. Short-term memory, also known as working memory, holds information for brief periods, allowing us to complete immediate tasks, such as calculating a tip or recalling a phone number long enough to dial it. Long-term memory, by contrast, involves the stabilization of information over time, a process heavily reliant on the hippocampus and surrounding medial temporal lobe structures. During consolidation, neural connections are strengthened through a process known as long-term potentiation, in which repeated activation of synapses enhances their efficiency. Over time, some memories are transferred from the hippocampus to the neocortex, where they become more stable and less dependent on the hippocampus for retrieval.

The storage of long-term memories is not uniform but instead categorized based on content. Explicit memories, also known as declarative memories, encompass facts and events that can be consciously recalled. These are further divided into episodic memories, which pertain to personal experiences, and semantic memories, which involve general knowledge about the world. Implicit memories, on the other hand, operate unconsciously and include skills and habits, such as riding a bike or typing on a keyboard. These are stored in distinct brain regions like the basal ganglia and cerebellum, highlighting the specialized nature of memory storage.

Retrieval, the process of accessing stored information, relies on the reactivation of neural circuits involved in the original encoding. When you try to remember a friend's birthday, for instance, your brain reconstructs the memory by piecing together fragments stored across various regions. This reconstruction is not always perfect, as memories are susceptible to distortion over time. Factors such as context, emotional state, and external cues play a significant role in retrieval, often shaping how and what we remember. A familiar scent or song, for instance, can trigger a cascade of memories, vividly transporting you to a specific moment in the past.

Emotion exerts a profound influence on memory formation and retrieval. Highly emotional experiences, whether positive or negative, tend to be remembered more vividly than neutral events. This heightened retention is mediated by the amygdala, a brain structure involved in processing emotions. The amygdala works in concert with the hippocampus, enhancing the encoding and consolidation of memories associated with strong emotions. This is why significant life events, such as weddings or accidents, often leave lasting impressions. However, the same mechanisms can also underlie intrusive memories in conditions like post-traumatic stress disorder, where emotionally charged recollections are repeatedly and involuntarily retrieved.

The brain's ability to form and retrieve memories is not static but evolves over the course of a lifetime. During childhood and adolescence, the developing brain is particularly adept at encoding and storing new information, thanks to heightened neural plasticity. As we age, however, changes in brain structure and function can impact memory. The hippocampus, for instance, tends to shrink with age, leading to difficulties in forming new memories. Despite these challenges, older adults often retain their ability to retrieve well-established memories, demonstrating the resilience of long-term storage.

Sleep plays a crucial role in memory consolidation and retrieval. During sleep, particularly during slow-wave and REM stages, the brain reactivates and reorganizes neural connections formed during wakefulness. This process strengthens newly acquired memories and integrates them with existing knowledge, enhancing recall and understanding. Sleep deprivation, on the other hand, can significantly impair memory, underscoring the importance of adequate rest for cognitive health.

Disruptions in memory formation and retrieval can have profound effects on daily life. Neurological conditions such as Alzheimer's disease, traumatic brain injuries, and stroke can impair the brain's ability to encode, store, or access information. In Alzheimer's, for example, the progressive accumulation of amyloid plaques and tau tangles interferes with synaptic function, particularly in the hippocampus, leading to memory loss and cognitive decline. Similarly, damage to specific brain regions can result in amnesia, a condition marked by the inability to form new memories (anterograde amnesia) or recall past ones (retrograde amnesia).

Despite these challenges, the brain's remarkable plasticity offers hope for mitigating memory impairments. Cognitive training, physical exercise, and even dietary interventions have been shown to support memory function by promoting neural health and enhancing synaptic connectivity. Advances in neuroscience have also led to the development of novel therapies, such as deep brain stimulation and pharmacological interventions, aimed at restoring memory in individuals with neurological disorders.

The intricate dance of encoding, storage, and retrieval that underpins memory is a cornerstone of human experience. It allows us to learn from the past, navigate the present, and plan for the future, weaving the fabric of our identities and relationships. While memory is not infallible and is subject to distortion and decay, its adaptability and resilience are a testament to the brain's extraordinary capabilities. By understanding the mechanisms that govern memory, we gain insight into the essence of what it means to be human, as well as the potential to enhance and protect this vital function.

Decision Making Pathways

Decision-making is one of the most intricate processes the human brain undertakes, requiring the integration of sensory information, memories, emotions, and predictions about the future. Every decision, from the mundane choice of what to eat for breakfast to the life-altering determination of a career path, involves a complex network of neural pathways working in concert. These pathways are not linear but deeply interconnected, reflecting the dynamic nature of human cognition and the necessity for balance between speed and deliberation.

At the heart of decision-making lies the prefrontal cortex, a region of the brain responsible for higher-order cognitive processes. This area, located just behind the forehead, is often regarded as the command center for decision-making. It evaluates options, weighs consequences, and selects actions based on goals and priorities. Within the prefrontal cortex, distinct subdivisions take on specialized roles. The dorsolateral prefrontal cortex, for instance, is heavily involved in rational, goal-directed decision-making. When you analyze the pros and cons of a potential choice, this region is actively working to ensure logical evaluation. In contrast, the ventromedial prefrontal cortex integrates emotional and social information, influencing decisions shaped by personal values, empathy, or a sense of reward. Together, these regions create a balance between objective reasoning and subjective personal context.

The limbic system, particularly the amygdala, also plays a crucial role in decision-making, especially under conditions of uncertainty or emotional intensity. The amygdala processes emotions such as fear, excitement, and anxiety, which can heavily influence choices. Imagine standing at the edge of a cliff and deciding whether to jump into the water below; the amygdala's assessment of risk and fear will likely shape your final choice. While the prefrontal cortex may calculate the safety of the jump based on logic, the amygdala's emotional input ensures that your decision is not devoid of instinctual caution. This interplay between rational and emotional systems highlights the dual nature of decision-making: a delicate balance between logic and intuition.

Another central player in decision-making is the basal ganglia, a group of structures involved in habit formation, reward processing, and action selection. The basal ganglia work closely with the dopaminergic system, which governs the release of dopamine, a neurotransmitter associated with reward and motivation. When faced with choices, the basal ganglia assess the potential rewards and risks of each option, often driving decisions based on past experiences. For example, if eating a particular food once provided a pleasurable experience, the basal ganglia may bias future decisions toward choosing that same food. This reward-based learning is critical for survival, encouraging behaviors that promote well-being and discouraging those that lead to harm.

During decision-making, different types of choices activate distinct neural circuits. Routine decisions, such as brushing your teeth or taking the usual route to work, rely heavily on the basal ganglia's ability to automate behaviors. These decisions require minimal cognitive effort, freeing up the prefrontal cortex for more demanding tasks. However, novel or complex decisions—those that involve uncertainty or significant consequences—engage the prefrontal cortex more extensively. In these cases, the brain must integrate information from multiple sources, including memory, sensory input, and emotional states, to arrive at a well-reasoned conclusion.

The anterior cingulate cortex, located deep within the brain, acts as a bridge between cognitive and emotional decision-making processes. This region monitors conflicts between competing options and evaluates the effort required to pursue a particular course of action. When you hesitate between two equally appealing choices, such as taking a promotion in a new city or staying close to family, the anterior cingulate cortex is actively resolving the internal conflict. It also plays a role in error detection, helping you learn from past mistakes by signaling when a decision leads to an unfavorable outcome.

Time pressure and stress can significantly alter decision-making pathways. When forced to make rapid decisions, the brain often bypasses the deliberate reasoning of the prefrontal cortex and relies instead on faster, more instinctual responses from the limbic system. This shift can be advantageous in high-stakes situations, such as avoiding an oncoming car, where immediate action is required. However, it can also lead to impulsive or suboptimal choices when there is insufficient time to evaluate all available information.

Social and environmental factors add yet another layer of complexity to decision-making. The orbitofrontal cortex, part of the prefrontal cortex, is particularly attuned to social cues and norms, influencing decisions within interpersonal contexts. For example, choosing how to respond during a heated argument may involve weighing the potential social consequences of your words. The influence of culture, upbringing, and peer pressure can shape decision-making pathways, often steering choices toward conformity or rebellion depending on the context.

One of the fascinating aspects of decision-making is the brain's ability to adjust pathways based on feedback and experience. This adaptability is a result of neuroplasticity, which allows neural connections to strengthen with repetition or weaken when no longer relevant. Consider learning to play a musical instrument: at first, every decision about where to place your fingers or how to read a note requires conscious effort. Over time, as the brain refines its pathways, these decisions become automatic, allowing for fluid and effortless performance. This same principle applies to decision-making in other areas of life, whether it's mastering a skill, navigating moral dilemmas, or refining personal habits.

Despite the brain's remarkable capacity for decision-making, biases and errors are inevitable. Cognitive biases, such as confirmation bias or loss aversion, can distort the evaluation of information, leading to flawed conclusions. These biases often arise because the brain relies on heuristics—mental shortcuts designed to simplify complex decisions. While heuristics can be efficient, they are not always accurate, particularly in unfamiliar or ambiguous situations. Recognizing and mitigating these biases is essential for improving decision-making, whether in personal life, professional settings, or broader societal contexts.

The pathways involved in decision-making are a testament to the brain's complexity and adaptability. They enable humans to navigate an unpredictable world, balancing immediate needs with long-term goals, rationality with emotion, and individuality with social dynamics. By understanding the mechanisms that underlie our choices, we gain insight into our behavior, enhance our ability to make informed decisions, and cultivate greater awareness of the factors that shape our lives. Each decision, no matter how small, reflects the intricate workings of a brain designed to adapt, learn, and thrive.

Chapter 3: Emotional Intelligence and Neural Correlates

The Limbic System Architecture

The limbic system is one of the most fascinating and integral structures of the brain, serving as the bridge between instinctual behaviors and higher cognitive functions. Nestled deep within the brain, it operates as the hub for emotion, memory, and motivation, influencing nearly every aspect of human experience. Its architecture, though complex, reflects the evolutionary history of the brain, with ancient structures intertwined with more recently developed regions. Understanding this intricate system provides insight into how we process emotions, form memories, and navigate the world through motivation and survival instincts.

The limbic system is not a single structure but a network of interconnected regions. At its core lies the hippocampus, a seahorse-shaped structure crucial for memory formation and spatial navigation. The hippocampus acts as the brain's archivist, transforming short-term memories into long-term ones and storing information about our surroundings. Without it, the ability to form new memories would be severely impaired, as seen in cases of individuals with hippocampal damage who experience profound anterograde amnesia. Beyond memory, the hippocampus also helps us orient ourselves in space, creating cognitive maps that allow us to navigate familiar and unfamiliar environments.

The amygdala, almond-shaped and situated near the hippocampus, is the emotional center of the limbic system. It processes emotions such as fear, anger, and pleasure, acting as an alarm system that alerts the brain to potential threats or rewards. When you feel a rush of fear while encountering something dangerous, it is the amygdala that has triggered this reaction. It plays a pivotal role in survival, ensuring that we respond appropriately to challenges in our environment. The amygdala also influences emotional memory, attaching significance to events and ensuring that those with emotional weight are more vividly remembered. This explains why moments of intense joy or trauma are often etched into our memories with remarkable clarity.

The hypothalamus, though small in size, wields immense influence over the body's internal state. Positioned just below the thalamus, it acts as the command center for maintaining homeostasis, regulating processes such as hunger, thirst, body temperature, and circadian rhythms. It serves as a link between the nervous system and the endocrine system, orchestrating the release of hormones that impact everything from stress responses to sexual behavior. The hypothalamus ensures that the body's needs are met, aligning physiological states with emotional and behavioral responses. For instance, when you experience hunger, the hypothalamus not only triggers the physical sensation but also motivates you to seek food.

The cingulate gyrus, a curved structure that arches over the corpus callosum, integrates emotional and cognitive information. It plays a role in regulating behavior and decision-making, particularly in situations involving conflict or uncertainty. When grappling with a difficult decision, the cingulate gyrus helps balance emotional impulses with rational thought, guiding you toward an appropriate course of action. It also contributes to social behavior, empathy, and the processing of pain, both physical and emotional.

The thalamus acts as the gateway between the sensory systems and the limbic system, relaying information from the senses to the appropriate cortical areas for processing. Though not exclusively part of the limbic system, its role is critical in shaping emotional and motivational responses to sensory stimuli. For example, a pleasant scent might trigger memories of a loved one or feelings of comfort, a connection made possible by the thalamus acting in concert with the hippocampus and amygdala.

The limbic system also includes structures such as the septal nuclei, which are associated with pleasure and reward, and the mammillary bodies, which are involved in memory processing. Together, these interconnected regions form a dynamic network that influences nearly every facet of human experience. While each component has its distinct role, their true power lies in their integration, as they work together to create a seamless interplay of emotion, memory, and motivation.

One of the defining features of the limbic system is its connection to the prefrontal cortex, the brain's center for higher-order thinking and executive function. This connection allows the limbic system to influence decision-making and behavior, ensuring that emotional and instinctual impulses are tempered by rational thought. However, this balance is not always perfect. Under conditions of stress or intense emotion, the limbic system can overpower the prefrontal cortex, leading to impulsive or irrational behavior. For instance, in moments of fear or anger, the amygdala may trigger a fight-or-flight response before the prefrontal cortex has had a chance to assess the situation fully.

The limbic system's architecture also reflects its evolutionary roots. Many of its components, such as the amygdala and hypothalamus, are ancient structures shared with other vertebrates, underscoring their importance for survival. These regions govern basic instincts such as feeding, reproduction, and self-preservation. The hippocampus and cingulate gyrus, while still ancient, are more developed in mammals, supporting the complex social behaviors and memory capabilities that characterize this group. This layering of old and new structures demonstrates how evolution has preserved and expanded upon the brain's foundational functions, enabling the emergence of higher cognition while retaining essential survival mechanisms.

Damage or dysfunction within the limbic system can have profound effects on behavior and cognition. Lesions in the hippocampus may result in memory impairments, while abnormalities in the amygdala are linked to anxiety disorders and phobias. Overactivity in the hypothalamus can disrupt hormonal balance, leading to issues such as chronic stress or metabolic disorders. The cingulate gyrus, when impaired, can contribute to conditions such as obsessive-compulsive disorder or depression. These examples highlight the delicate balance maintained by the limbic system and its profound influence on mental and physical health.

Despite its complexity, the limbic system is highly adaptable, capable of rewiring itself in response to experience and learning. This plasticity allows it to adjust to changing circumstances, strengthening connections that support positive behaviors and weakening those associated with negative ones. For example, exposure therapy for phobias capitalizes on the brain's ability to reduce the amygdala's response to fear-inducing stimuli through repeated exposure in a controlled environment. Similarly, mindfulness practices and cognitive-behavioral therapy can enhance the connection between the prefrontal cortex and the limbic system, promoting greater emotional regulation and resilience.

The architecture of the limbic system serves as a foundation for understanding the intricate interplay between emotion, memory, and motivation. Its components, both ancient and modern, work together to shape our experiences and responses, reflecting the remarkable complexity and adaptability of the human brain. By examining this system, we gain insight into how we navigate the world, connect with others, and create meaning from our experiences, revealing the

profound role this network plays in defining who we
are.

Emotion Cognition Interaction

The intricate dance between emotion and cognition is
a cornerstone of human experience, shaping
decisions, behaviors, and the way individuals perceive
the world around them. Emotions, often regarded as
fleeting and sometimes irrational, are deeply
entwined with the cognitive processes that govern
thought, memory, and reasoning. This interaction is
not merely a parallel occurrence but a dynamic
relationship where each influences and modulates the
other.

When faced with a situation, emotions often serve as
the first responders, providing an immediate, visceral
reaction that precedes deliberate thought. This initial
emotional response stems from the brain's limbic
system, particularly the amygdala, which processes
stimuli and triggers reactions rooted in survival
instincts. Fear, for instance, prepares the body to flee
or fight, while joy may encourage connection and
engagement. These emotional signals act as a
guidepost, directing attention and prioritizing certain
stimuli over others. Without this emotional filter, the
overwhelming influx of information from the
environment would be almost impossible to navigate
efficiently.

Cognition, however, plays a critical role in
interpreting and regulating these emotional
responses. The prefrontal cortex, responsible for

executive functions such as decision-making and impulse control, works to assess the context of an emotional reaction. It allows individuals to evaluate whether a response is appropriate or whether it requires adjustment. For instance, while a sudden loud noise might initially trigger fear, cognitive evaluation would determine whether the noise poses an actual threat or is benign, such as the sound of fireworks. This interplay ensures that emotions are not only reactive but also adaptive, aligning with the complexities of modern life.

Memory serves as a bridge between emotion and cognition, further solidifying their interconnectedness. Emotional experiences are often encoded more vividly and are easier to recall than neutral ones. This phenomenon is due to the activation of the amygdala during emotionally charged events, which enhances the consolidation of memories in the hippocampus. A tragic event, a moment of triumph, or the warmth of a cherished memory becomes etched in the mind, influencing future decisions and behaviors. These memories, in turn, provide a cognitive framework through which similar situations are assessed, drawing on past emotions to guide present actions.

Decision-making is another domain where the interaction between emotion and cognition becomes particularly evident. While traditional models of rationality suggest that decisions should be made solely based on logic and reason, real-life choices are rarely devoid of emotional influence. Emotions provide a shortcut, a heuristic that simplifies complex decision-making processes. For example, a feeling of unease about a particular option might stem from past experiences or subconscious recognition of potential risks. Conversely, a sense of excitement may drive one toward opportunities that align with personal values and desires. This emotional input, when balanced with cognitive deliberation, often leads to more holistic and effective decisions.

There are instances, however, where the balance between emotion and cognition can become disrupted, leading to maladaptive outcomes. Intense emotions, such as anger or despair, can overshadow rational thought, resulting in impulsive actions or distorted perceptions. Similarly, an overreliance on cognitive processes while suppressing emotions can lead to rigid thinking and a lack of empathy. Striking a balance between these two forces is essential for psychological well-being, fostering resilience, adaptability, and a deeper understanding of oneself and others.

Cultural and individual differences also play a significant role in shaping the interaction between emotion and cognition. In some cultures, emotional expression is encouraged and integrated into decision-making processes, while in others, restraint and rationality are valued more highly. On an individual level, factors such as personality, upbringing, and life experiences influence how emotions and cognition interact. For instance, someone with a history of trauma may have heightened emotional responses that require greater cognitive effort to manage, while another individual might naturally lean toward a more analytical approach.

Modern neuroscience and psychology have shed light on practical ways to harness the power of emotion-cognition interaction. Mindfulness practices, for example, cultivate awareness of emotional states without judgment, allowing for a more measured cognitive response. This approach helps to break the cycle of reactive behavior, fostering a sense of control and intentionality. Similarly, emotional intelligence—the ability to recognize, understand, and manage emotions—enhances the interplay between emotion and cognition, enabling more effective communication, conflict resolution, and decision-making.

The dynamic between emotion and cognition is not a static relationship but rather a fluid and evolving process. Life experiences, learning, and personal growth continuously reshape the ways in which these two forces interact. By understanding and embracing this interplay, individuals can navigate the complexities of life with greater clarity and purpose. Recognizing the value of both emotional intuition and cognitive analysis allows for a more balanced perspective, one that honors the richness of human experience while promoting thoughtful and intentional action.

Social Brain Networks

The human brain is an intricate web of neural connections that not only governs individual cognition and behavior but also enables the deeply social nature of the human species. Social brain networks are the specialized systems that allow individuals to navigate the complexity of social interactions, interpret the behaviors and intentions of others, and build relationships. These networks are not isolated to one region but span multiple interconnected areas of the brain, each contributing to the understanding and engagement with the social world.

At the heart of social cognition lies the ability to infer the thoughts, emotions, and intentions of others, a skill often referred to as theory of mind. The medial prefrontal cortex is a critical player in this process, aiding in the interpretation of subtle social cues and the formation of mental models about other people's perspectives. This region works in tandem with the temporoparietal junction, which helps distinguish between one's own thoughts and those of others. Together, they create a framework for empathy, allowing individuals to step into another's shoes and respond with sensitivity and understanding.

Another vital component of social brain networks is the amygdala, a structure known for its role in processing emotions. In social contexts, the amygdala helps evaluate the emotional significance of facial expressions, tone of voice, and other nonverbal cues. It is particularly active during moments of heightened emotional interaction, such as recognizing fear or anger in another person. This rapid assessment of emotional states enables individuals to adapt their responses accordingly, fostering smoother and more effective communication.

The inferior frontal gyrus and its mirror neuron system also play a crucial role in social brain networks. Mirror neurons, first discovered in primates, fire both when an individual performs an action and when they observe someone else performing the same action. This mirroring mechanism facilitates imitation, learning, and the intuitive understanding of others' behaviors. For example, watching someone smile often triggers a smile in return, creating a shared emotional experience and reinforcing social bonds. These neurons are thought to be fundamental in developing empathy and understanding, as they allow one person to feel as though they are experiencing the emotions or actions of another.

Another key aspect of social brain networks involves the default mode network, a system of brain regions that becomes active during introspective and self-referential thought. This network, which includes the medial prefrontal cortex, posterior cingulate cortex, and angular gyrus, supports the ability to reflect on personal experiences and relate them to others. It enables individuals to consider how their actions and words might be perceived, fostering self-awareness and social sensitivity. This introspective capacity is essential for maintaining relationships, resolving conflicts, and understanding the nuances of complex social dynamics.

Social brain networks are not static; they are highly adaptable and shaped by experience. From infancy, interactions with caregivers and peers sculpt these neural systems, laying the foundation for future social skills. A baby's gaze into its caregiver's eyes, the reciprocal exchange of smiles, and the soothing tone of a parent's voice are all moments that activate and strengthen these networks. Over time, these interactions build the neural architecture necessary for more sophisticated social abilities, such as collaboration, negotiation, and emotional regulation.

The development and functioning of social brain networks can be influenced by a variety of factors, including genetics, environment, and individual experiences. Conditions such as autism spectrum disorder (ASD) and social anxiety disorder highlight the importance of these networks, as they often involve differences or disruptions in social processing. For example, individuals with ASD may have difficulty interpreting facial expressions or understanding the perspectives of others, reflecting differences in how their social brain networks operate. Meanwhile, social anxiety can amplify the activity of the amygdala, leading to heightened sensitivity to perceived social threats and challenges in navigating social situations.

Cultural context also plays a significant role in shaping social brain networks. Different cultures emphasize distinct social norms, values, and patterns of interaction, which can influence how these networks are engaged and developed. For instance, cultures that prioritize collectivism may foster stronger neural connections related to group harmony and interdependence, while individualistic cultures may emphasize self-expression and personal achievement. This cultural variability demonstrates the brain's remarkable plasticity and its ability to adapt to diverse social environments.

Technology and modern communication methods have introduced new dimensions to the functioning of social brain networks. Virtual interactions, such as video calls or social media exchanges, require the brain to interpret and respond to social cues in novel ways. While these technologies can enhance connectivity and broaden social networks, they also present challenges, such as the potential for misinterpreting text-based communication or the diminished impact of nonverbal cues. Understanding how social brain networks adapt to these changes offers valuable insight into the evolving nature of human interaction.

The study of social brain networks sheds light on what it means to be human. These systems are not only responsible for individual social abilities but also underpin the cooperative behaviors that have allowed humans to thrive as a species. Through collaboration, empathy, and shared understanding, individuals form communities, build civilizations, and create the rich tapestry of human culture. By better understanding the neural mechanisms that drive these interactions, it becomes possible to cultivate stronger relationships, foster inclusivity, and address challenges that arise in social contexts. Social brain networks are a testament to the profound interconnectedness of human lives, bridging the gap between individual minds and the collective experience of humanity.

Empathy and Mirror Neurons

Empathy is one of the most profound and defining traits of humanity, allowing individuals to connect with one another on emotional, cognitive, and experiential levels. It is the ability to understand and share the feelings of others, to bridge the gap between separate minds and create a shared emotional landscape. While empathy has long been considered a cornerstone of human interaction, neuroscience has uncovered a key mechanism that underlies this capacity: mirror neurons. These specialized cells, first identified in the early 1990s, provide a neural foundation for understanding and mimicking the actions and emotions of those around us, offering remarkable insight into how empathy operates in the brain.

Mirror neurons were first discovered in the premotor cortex of macaque monkeys, where researchers observed that specific neurons fired not only when the monkey performed an action, such as grasping an object, but also when it observed another monkey or a human performing the same action. This mirroring response suggested a neural system designed to map the actions of others onto one's own motor framework, creating a shared understanding of behavior. Subsequent research revealed that humans possess a similar system, with mirror neurons distributed across areas such as the inferior frontal gyrus, inferior parietal lobule, and supplementary motor areas. What began as an investigation into action recognition evolved into a deeper exploration of how these neurons contribute to a range of social and emotional processes, including empathy.

The role of mirror neurons in empathy is rooted in their ability to bridge the gap between observation and experience. When you see someone experiencing a strong emotion, such as joy or sadness, your mirror neurons activate as if you were experiencing that same emotion yourself. This neural mirroring helps to create an internal representation of the other person's state, allowing you to feel a version of what they are feeling. For instance, watching a friend's face light up with happiness can evoke a sense of shared joy within you, while witnessing someone in distress might trigger feelings of compassion or sorrow. This automatic and unconscious process forms the basis of emotional empathy, the capacity to feel with someone else.

Beyond emotional resonance, mirror neurons also play a role in cognitive empathy, which involves understanding another person's perspective or mental state. By simulating the observed actions, expressions, and behaviors of others, these neurons provide a framework for interpreting their intentions and emotions. For example, seeing someone slump their shoulders and avert their gaze might lead your mirror system to activate a similar posture in your mind, helping you infer that the person is feeling defeated or withdrawn. This ability to mentally step into someone else's experience is crucial for navigating social interactions, fostering cooperation, and resolving conflicts.

However, the functioning of empathy and mirror neurons is not a one-size-fits-all phenomenon. Individual differences, shaped by factors such as genetics, upbringing, and personal experiences, influence how strongly these systems operate. Some people may have a heightened sensitivity to the emotions of others, while others may struggle to accurately interpret social cues. Disorders such as autism spectrum disorder (ASD) have been linked to

differences in mirror neuron activity, which may
contribute to challenges in social communication and
empathy. Similarly, conditions like psychopathy or
certain personality disorders are associated with
deficits in empathic processing, raising questions
about how the mirror system interacts with other
neural and psychological mechanisms.

Context also plays a significant role in shaping
empathy. The brain's response to the emotions of
others is not purely automatic; it is influenced by
factors such as familiarity, cultural norms, and in-
group or out-group dynamics. People are generally
more likely to empathize with those they perceive as
similar to themselves, whether in terms of shared
identity, values, or experiences. This bias, while
rooted in evolutionary survival strategies, can also
create barriers to broader social understanding and
compassion. Overcoming such limitations requires
deliberate effort to expand one's empathic reach,
challenging ingrained assumptions and cultivating a
willingness to connect with diverse perspectives.

The plasticity of the brain offers hope for enhancing
empathy. While mirror neurons provide the
foundation, empathy itself can be cultivated through
conscious practice and learning. Activities such as
mindfulness meditation, active listening, and
exposure to different cultures and viewpoints have
been shown to strengthen empathic abilities. For
instance, mindfulness encourages individuals to pay
attention to their own emotions and those of others
without judgment, fostering a deeper awareness of the
shared human experience. Similarly, engaging with
literature, art, and storytelling can activate the mirror

neuron system by immersing the observer in the emotions and actions of fictional characters, providing a safe space to practice empathic understanding.

The implications of empathy and mirror neurons extend far beyond individual relationships. They form the basis for larger societal dynamics, influencing everything from education and healthcare to conflict resolution and policymaking. In classrooms, teachers who model empathic behavior can create supportive environments that enhance student learning and emotional well-being. In clinical settings, healthcare providers who demonstrate empathy can build trust and improve patient outcomes. On a broader scale, fostering empathy within communities can bridge divides, reduce prejudice, and promote social cohesion.

Empathy and the neural mechanisms that support it are a testament to the interconnectedness of human beings. Mirror neurons reveal that the boundaries between self and other are not as rigid as they may seem, offering a glimpse into the shared neural architecture that underpins social life. By understanding and nurturing these systems, individuals can deepen their capacity for connection, creating a ripple effect that extends through families, communities, and beyond. The study of empathy and mirror neurons not only illuminates the workings of the brain but also underscores the potential for humans to transcend isolation and build a more compassionate world.

Emotional Regulation Mechanisms

Emotional regulation is a complex and essential process that allows individuals to manage their feelings, reactions, and behaviors in response to life's challenges and experiences. It is not simply about suppressing emotions but rather understanding, modulating, and expressing them in ways that align with personal goals, social norms, and situational demands. This intricate ability is underpinned by a network of biological, cognitive, and social mechanisms, each working together to maintain emotional balance and resilience.

The brain plays a central role in emotional regulation, with various regions contributing to the modulation of emotional responses. The prefrontal cortex, particularly the ventromedial and dorsolateral areas, is a key player in this process. It is responsible for executive functions such as decision-making, impulse control, and attention, all of which are critical for regulating emotions. By assessing the context of an emotional event, the prefrontal cortex helps to determine an appropriate response, whether it involves calming oneself, redirecting attention, or reevaluating a situation. For instance, when faced with a frustrating scenario, such as a long traffic jam, the prefrontal cortex might help shift focus away from the inconvenience and toward more constructive thoughts, like using the time to listen to a favorite podcast.

The amygdala, a small almond-shaped structure in the brain, is another crucial component of emotional regulation. It acts as a sort of alarm system, detecting and processing emotional stimuli, particularly those related to fear and threat. While the amygdala's rapid response is essential for survival, such as reacting to danger, its activity must be balanced by the prefrontal cortex to prevent overreactions or prolonged emotional distress. This dynamic interplay between the amygdala and prefrontal cortex ensures that emotions are appropriately regulated, allowing for both quick responses to immediate challenges and thoughtful reflection in less urgent situations.

One of the most effective emotional regulation strategies involves cognitive reappraisal, the process of reinterpreting a situation to change its emotional impact. This mechanism relies on the brain's ability to shift perspective, reframing events in a way that reduces negative emotions or enhances positive ones. For example, instead of viewing a failed project as a personal failure, one might reframe it as a valuable learning experience that contributes to growth and future success. Cognitive reappraisal not only decreases stress and anxiety but also fosters a greater sense of control over emotional experiences.

Another important mechanism for regulating emotions is attentional control, which involves directing focus away from distressing stimuli and toward more neutral or positive aspects of a situation. This can be as simple as deliberately choosing to concentrate on a calming image, engaging in a distracting activity, or actively seeking out positive experiences. By shifting attention, individuals can reduce the intensity of negative emotions and gain the mental space needed to process challenging situations more effectively. For instance, someone overwhelmed by a heated argument might step away to focus on their breathing or immerse themselves in a hobby, allowing emotions to settle before revisiting the issue.

Behavioral strategies also play a significant role in emotional regulation. Physical activities such as exercise, yoga, or even a brisk walk can help release built-up tension and improve mood by triggering the release of endorphins and reducing stress hormones like cortisol. Similarly, social interactions provide a valuable outlet for emotional expression and support. Talking to a trusted friend, family member, or therapist can help process emotions, gain perspective, and feel less isolated in the face of difficulties.

Cultural and social influences shape emotional regulation mechanisms in profound ways. Cultural norms dictate how emotions should be expressed or restrained, creating frameworks within which individuals learn to manage their feelings. In some cultures, open emotional expression is encouraged, while in others, restraint and composure are valued more highly. These cultural differences influence which strategies individuals rely upon and how they interpret their own emotional experiences. Socialization, particularly during childhood, also plays a critical role. Children learn emotional regulation skills through observing and mimicking caregivers, receiving guidance on appropriate responses, and encountering various emotional scenarios in social settings. Over time, these lessons become internalized, forming the foundation for lifelong emotional regulation abilities.

Despite the many mechanisms that facilitate emotional regulation, challenges can arise when these systems are disrupted or underdeveloped. Stress, trauma, mental health disorders, and even temporary factors like fatigue can impair the brain's ability to modulate emotions effectively. For instance, heightened activity in the amygdala combined with reduced prefrontal cortex engagement can lead to heightened emotional reactivity and difficulty calming down after upsetting events. Chronic dysregulation can contribute to conditions such as anxiety, depression, or impulsive behaviors, further complicating the ability to manage emotions.

Building and strengthening emotional regulation skills is possible through deliberate practice and intervention. Mindfulness techniques, which emphasize present-moment awareness and nonjudgmental acceptance of emotions, have proven to be highly effective. By cultivating an awareness of emotional states without becoming overwhelmed by them, individuals can better navigate their feelings and respond in healthier ways. Similarly, practices such as journaling, deep breathing, and progressive muscle relaxation offer practical tools for calming the mind and body during moments of distress.

Therapeutic approaches, such as cognitive-behavioral therapy (CBT), also provide valuable support for developing emotional regulation. CBT helps individuals identify negative thought patterns, challenge irrational beliefs, and replace them with more constructive perspectives. By addressing the cognitive roots of emotional dysregulation, therapy fosters greater self-awareness and equips individuals with strategies for managing even the most intense emotions. For those facing severe challenges, medication or other medical interventions may be necessary to restore balance and support the brain's natural regulatory mechanisms.

Emotional regulation is not a static trait but a dynamic skill that evolves over time and through experience. It is the interplay of biology, cognition, and environment that enables individuals to navigate the highs and lows of life with resilience and adaptability. By understanding the mechanisms that underlie emotional regulation, individuals can cultivate greater self-mastery, improve their relationships, and enhance their capacity to thrive in an ever-changing world. The ability to regulate emotions is a testament to the remarkable complexity of the human mind, reflecting both its vulnerabilities and its extraordinary potential.

Chapter 4: Learning and Memory Systems

Sensory Processing and Storage

The human experience is shaped by the constant influx of sensory information from the environment, a stream of stimuli that the brain processes, interprets, and stores to create a coherent understanding of the world. Sensory processing and storage involve a sophisticated series of neural mechanisms that begin with the detection of external stimuli and culminate in the formation of memories. This intricate process not only allows individuals to navigate their surroundings but also plays a crucial role in learning, decision-making, and emotional responses.

Sensory processing begins with the specialized sensory organs—eyes, ears, skin, nose, and tongue—which serve as the gateways between the external world and the nervous system. Each organ is equipped with specific receptors that detect particular types of stimuli. Photoreceptors in the retina of the eye respond to light, mechanoreceptors in the skin detect pressure and vibration, and hair cells in the cochlea of the ear pick up sound waves. These receptors convert physical stimuli into electrical signals through a process known as transduction, enabling the brain to interpret sensory input. Once converted, these signals travel via specialized neural pathways to the brain, where they are further processed and integrated.

Sensory information first reaches the thalamus, often referred to as the brain's relay station. The thalamus acts as a hub, directing sensory data to the appropriate cortical areas for further processing. Visual stimuli are sent to the occipital lobe, auditory signals to the temporal lobe, and tactile information to the somatosensory cortex in the parietal lobe. Each of these areas specializes in analyzing specific types of sensory input, breaking down stimuli into their constituent parts. For instance, the visual cortex identifies elements such as color, shape, and motion, while the auditory cortex deciphers pitch, tone, and rhythm. This specialized processing is essential for extracting meaningful patterns from the raw data provided by sensory organs.

Integration occurs once the brain has processed individual sensory modalities. Multisensory integration allows the brain to combine input from different senses to create a unified perception of the environment. For example, hearing the sound of a bird chirping while simultaneously seeing its movement through the trees enables a cohesive understanding of the scene. This integration is facilitated by regions such as the superior colliculus and the temporoparietal junction, which merge sensory data into a single, comprehensive representation. Without this ability, the world would appear fragmented, making it difficult to interpret complex situations or respond effectively to environmental demands.

As sensory input is processed, the brain determines which information is relevant for immediate use and which should be stored for future reference. Attention plays a critical role in this selection process, acting as a filter that prioritizes certain stimuli over others. Novel or emotionally significant stimuli are more likely to capture attention and be encoded into memory. For instance, the scent of a specific perfume might stand out and be remembered because it evokes a strong emotional connection to a loved one. Conversely, mundane or repetitive stimuli, such as the faint hum of an air conditioner, may fade into the background and be disregarded.

The transformation of sensory input into memory begins with encoding, the process by which sensory data is converted into a format that can be stored in the brain. This involves the activation of neural circuits in the hippocampus and surrounding medial temporal lobe structures, which play a pivotal role in consolidating sensory experiences into long-term memories. Encoding can be influenced by factors such as attention, emotional state, and prior knowledge, all of which enhance the likelihood that sensory information will be retained. For instance, an emotionally charged event, like a wedding or a traumatic accident, is more likely to be vividly encoded and stored than a routine task like grocery shopping.

Once encoded, sensory memories are stored in different regions of the brain depending on their type. Visual memories, for example, are largely stored in the visual cortex, while auditory memories are held in the auditory cortex. The prefrontal cortex is responsible for working memory, a short-term storage system that allows individuals to hold and manipulate sensory information for immediate use. Long-term memories, on the other hand, are distributed across a network of cortical areas, with the hippocampus coordinating their retrieval and integration. This distributed storage ensures that memories are robust and less vulnerable to damage or loss.

Sensory storage is not a static process; it is continuously updated and reshaped by new experiences and information. Reconsolidation, the process of reactivating and modifying stored memories, allows the brain to integrate new sensory input with existing knowledge. This dynamic capability is essential for learning and adaptation, enabling individuals to refine their understanding of the world over time. However, it also means that memories are malleable and subject to distortion. For example, recalling a past event may unintentionally incorporate details from subsequent experiences, altering the original memory.

Disruptions in sensory processing and storage can have profound effects on perception and behavior. Conditions such as sensory processing disorder (SPD) and post-traumatic stress disorder (PTSD) highlight the importance of these mechanisms. In SPD, individuals may experience hypersensitivity or hyposensitivity to sensory stimuli, leading to difficulties in daily functioning. In PTSD, traumatic sensory memories can become over-consolidated, resulting in intrusive recollections and heightened emotional responses. Understanding these disruptions provides valuable insight into the neural processes underlying sensory processing and storage, as well as potential avenues for intervention and treatment.

The remarkable efficiency and complexity of sensory processing and storage underscore their significance in shaping human experience. By converting external stimuli into meaningful perceptions and memories, these mechanisms form the foundation for everything from basic survival to the appreciation of art, music, and literature. They enable individuals to learn from the past, adapt to the present, and plan for the future, demonstrating the brain's extraordinary capacity to transform raw sensory data into the richness of conscious life. Sensory processing and storage not only connect individuals to their environment but also to one another, creating a shared reality built on the intricate interplay of perception and memory.

Learning Mechanisms at Neural Level

Learning is a fundamental process that enables humans to adapt, grow, and respond to an ever-changing environment. At its core, learning is deeply rooted in the neural architecture of the brain, where intricate mechanisms translate experiences into lasting changes in behavior, knowledge, and skills. These mechanisms operate on a cellular level, involving dynamic interactions between neurons, synapses, and networks that form the foundation of memory, adaptation, and cognitive development. Understanding learning at the neural level illuminates the profound complexity of this process and reveals how the brain encodes, stores, and retrieves information.

Neural learning begins with the brain's most basic unit: the neuron. Neurons communicate with one another through synapses, the tiny gaps that separate one neuron's axon terminal from another's dendrite. These synapses are the sites where chemical and electrical signals are exchanged, forming the basis for neural communication and the transmission of information. When an individual encounters new information or experiences, specific neural pathways are activated. These pathways become stronger with repeated use, a phenomenon known as synaptic plasticity. Synaptic plasticity is the brain's ability to modify the strength and efficiency of synaptic connections, a critical mechanism for learning and memory formation.

One of the most well-known forms of synaptic plasticity is long-term potentiation (LTP). LTP occurs when repeated stimulation of a synapse results in a sustained increase in the strength of the connection between two neurons. This process is often described as the cellular basis of learning and memory, as it allows neural circuits to encode and retain information over time. LTP primarily involves the release of neurotransmitters such as glutamate, which bind to receptors on the postsynaptic neuron. The activation of these receptors, particularly N-methyl-D-aspartate (NMDA) receptors, triggers a cascade of molecular events that enhance synaptic strength. Over time, this strengthening of synaptic connections solidifies the neural representation of learned information.

Conversely, long-term depression (LTD) serves as a complementary mechanism to LTP. LTD involves a decrease in the strength of synaptic connections, which occurs when synapses are activated less frequently or in a specific pattern. This weakening of synapses is essential for pruning redundant or irrelevant connections, clearing space for new learning and ensuring the efficiency of neural networks. The balance between LTP and LTD allows the brain to remain flexible and adaptable, enabling the continuous refinement of knowledge and skills.

Beyond synaptic plasticity, structural changes in the brain also contribute to learning. Neurogenesis, the process of generating new neurons, plays a role in specific forms of learning, particularly in the hippocampus, a region critical for memory and spatial navigation. While neurogenesis occurs predominantly during early development, evidence suggests that it continues into adulthood, albeit at a slower rate. These newly generated neurons integrate into existing neural circuits, enhancing the brain's capacity for adaptation and the formation of new memories. Additionally, the growth and remodeling of dendritic spines—tiny protrusions on neurons that receive synaptic inputs—further support learning by increasing the surface area available for synaptic connections.

Different types of learning rely on distinct neural circuits and mechanisms. Declarative learning, which involves acquiring facts and information, depends heavily on the medial temporal lobe, including the hippocampus. This region acts as a hub for consolidating new memories, transferring them from short-term to long-term storage in the cortical areas. Procedural learning, on the other hand, involves the acquisition of motor skills and habits and is mediated by the basal ganglia and cerebellum. These structures coordinate the fine-tuning of movements and the automation of repeated actions, enabling individuals to perform tasks with increasing accuracy and efficiency over time.

Emotional learning, which ties experiences to emotional responses, engages the amygdala, a region of the brain involved in processing emotions. When an event carries significant emotional weight, such as a moment of danger or triumph, the amygdala interacts with the hippocampus to enhance the encoding and retention of the associated memory. This interaction ensures that emotionally charged experiences are remembered more vividly, serving as a survival mechanism that helps individuals avoid threats or seek rewards in the future.

The brain's ability to learn is also influenced by neurotransmitters, chemicals that modulate neural activity and communication. Dopamine, for example, plays a crucial role in reward-based learning by signaling the anticipation and receipt of rewards. This neurotransmitter reinforces behaviors that lead to positive outcomes, strengthening the neural circuits associated with those behaviors. Acetylcholine, on the other hand, is involved in attention and the enhancement of synaptic plasticity, ensuring that the brain prioritizes and retains relevant information. These chemical messengers work in concert with neural circuits to optimize the learning process.

Learning at the neural level is not a static process; it is shaped by experience, environment, and individual differences. Factors such as stress, sleep, and nutrition can significantly impact neural plasticity and the brain's capacity to learn. Chronic stress, for instance, can impair the function of the hippocampus and prefrontal cortex, hindering memory formation and decision-making. Conversely, adequate sleep supports the consolidation of memories and the pruning of unnecessary synaptic connections. Environments rich in stimulation and novelty also enhance neural plasticity, promoting the growth of new connections and the reinforcement of existing ones.

While the brain's capacity for learning is remarkable, it is not without limits. Over time, synaptic connections can become saturated, leading to a plateau in learning. This phenomenon underscores the importance of spacing and variation in practice, which allow the brain to recover and reorganize before acquiring additional information. Furthermore, the brain's plasticity declines with age, making it more challenging to form new connections and retain information. However, engaging in lifelong learning and mentally stimulating activities can help mitigate these effects, preserving cognitive function and adaptability.

The neural mechanisms underlying learning reveal the extraordinary complexity and adaptability of the brain. From the strengthening of synaptic connections to the generation of new neurons, these processes form the foundation for acquiring knowledge, developing skills, and adapting to a dynamic world. By understanding how learning operates at the neural level, individuals can harness these mechanisms to optimize their own learning experiences, overcome challenges, and continue evolving throughout life. The brain's capacity to learn is a testament to its resilience and ingenuity, a reflection of the intricate interplay between biology and experience that defines human potential.

Memory Consolidation Pathways

Memory consolidation is the intricate process through which temporary, short-term memories are transformed into enduring, long-term ones. This dynamic operation occurs through a network of interconnected pathways in the brain, involving both structural and biochemical alterations. It is not a singular event but rather a series of stages influenced by neural activity, external stimuli, and the passage of time. The brain's ability to retain and organize information depends significantly on how well these pathways operate, shaping the foundation of learning, identity, and daily functionality.

The hippocampus plays a pivotal role in memory consolidation, serving as a central hub for the processing and storage of new information. When an experience or piece of knowledge is first encountered, the hippocampus acts as a temporary storage site, encoding the memory and linking it to relevant sensory and contextual data. This initial phase, known as encoding, relies heavily on electrical activity and the firing of neurons. Neural synapses, the connections between brain cells, are strengthened to encode this new information. However, memories do not remain within the hippocampus indefinitely. Over time, they are transferred to other regions of the brain, such as the neocortex, for long-term storage. This gradual transfer is essential for freeing up the hippocampus to process new experiences.

Consolidation occurs through two primary mechanisms: synaptic consolidation and systems consolidation. Synaptic consolidation takes place on a cellular level, occurring within hours of the initial memory formation. This process involves the stabilization of synaptic connections, primarily through the release and reception of neurotransmitters like glutamate, which bind to receptors and trigger a cascade of molecular changes. Proteins are synthesized, and synaptic connections are fortified, creating a more stable memory trace. Without this phase, memories remain fragile and susceptible to disruption. On the other hand, systems consolidation is a more prolonged process that can take days, weeks, or even years. This stage involves the reorganization of memory traces, as they are gradually distributed across various cortical regions. The repeated activation of these pathways strengthens the connections, embedding the memory into the brain's architecture.

Sleep plays a critical role in enhancing memory consolidation, acting as a natural facilitator of this intricate process. During sleep, particularly in the deep stages of slow-wave sleep, the brain engages in a unique pattern of activity. The hippocampus replays recently encoded memories, sending signals to the neocortex to reinforce and integrate them. This neural replay effectively reactivates the memory traces, strengthening their connections and ensuring their transfer to long-term storage. Meanwhile, rapid eye movement (REM) sleep is associated with the reorganization and integration of memories, particularly emotional ones. Dreams, often a byproduct of REM sleep, may serve as a mechanism for processing and linking disparate pieces of information, further aiding consolidation.

Stress and emotions also have a profound impact on memory consolidation, for better or worse. High-stress situations trigger the release of stress hormones, such as cortisol and adrenaline, which can either enhance or impair the consolidation process depending on the context and intensity. Moderate levels of stress can heighten attention and arousal, leading to stronger memory encoding and consolidation. For instance, emotionally charged events, whether positive or negative, are often remembered more vividly due to the heightened activity in the amygdala, a brain region closely tied to emotional processing. However, chronic stress or overwhelming emotional responses can disrupt the delicate balance of neurotransmitters and impair the hippocampus's functioning, leading to fragmented or weakened memories.

Physical activity and a healthy lifestyle can also influence the efficiency of memory consolidation pathways. Regular exercise has been shown to increase the production of brain-derived neurotrophic factor (BDNF), a protein that supports the growth and maintenance of neurons. BDNF enhances synaptic plasticity, allowing the brain to adapt and optimize memory pathways. Similarly, a balanced diet rich in omega-3 fatty acids, antioxidants, and essential vitamins provides the necessary nutrients for brain health. Adequate hydration and avoiding harmful substances such as excessive alcohol further contribute to the optimal functioning of memory consolidation mechanisms.

Disruptions to memory consolidation pathways can lead to significant cognitive challenges. Conditions such as Alzheimer's disease, traumatic brain injuries, and sleep disorders interfere with the brain's ability to organize and store memories effectively. In Alzheimer's disease, for example, the accumulation of amyloid plaques and tau tangles disrupts neural communication and impairs the consolidation process. Similarly, individuals with sleep apnea or chronic insomnia often experience difficulties in retaining new information due to the lack of restorative sleep. Understanding these disruptions provides insight into potential interventions, such as cognitive therapies, medications, or lifestyle modifications, to mitigate their impact.

The interplay between memory consolidation pathways is complex and constantly evolving. Advances in neuroscience continue to shed light on the intricate mechanisms underlying this process, paving the way for innovative approaches to enhance memory retention and retrieval. By recognizing the factors that influence consolidation, from sleep and stress to nutrition and neural plasticity, it becomes possible to optimize the brain's capacity to learn and remember. The journey of memory, from fleeting moments to enduring recollections, remains one of the most fascinating mysteries of the human mind, a testament to the brain's remarkable adaptability and resilience.

Neural Plasticity in Learning

Neural plasticity is the foundation of the brain's ability to adapt, learn, and reorganize itself in response to new information, experiences, and environmental changes. This remarkable capacity enables the brain to modify its structure and function throughout life, making it a dynamic organ capable of growth and transformation. Learning, in particular, is intricately tied to this adaptability, as the brain forms, strengthens, and refines connections between neurons to encode knowledge and skills. The concept of neural plasticity reveals how experiences can leave a lasting imprint on the brain, shaping not only cognition but also behavior and perception.

The process of learning begins when the brain encounters new stimuli, whether it's an unfamiliar concept, a skill to be mastered, or even a subtle change in the environment. When this happens, neurons communicate through electrical impulses and chemical signals, creating pathways that encode the new information. These pathways, known as neural circuits, are not fixed; instead, they are constantly being shaped and refined by neural plasticity. Synapses, the junctions between neurons, play a crucial role in this process. When a new memory or skill is formed, synaptic connections are either strengthened or weakened, depending on the frequency and intensity of their activation. The principle of "use it or lose it" governs this process, as frequently activated synapses grow stronger while unused ones are pruned away.

One of the most striking features of neural plasticity is its ability to create entirely new connections between neurons. This process, known as synaptogenesis, is particularly robust during early childhood, a period marked by rapid cognitive and emotional development. During this time, the brain is highly malleable, forming an abundance of synaptic connections to support learning and adaptation. However, neural plasticity is not confined to childhood; it continues throughout life, albeit at a slower pace. Adults retain the ability to learn and adapt because the brain remains capable of reorganizing itself in response to new challenges and experiences.

Long-term potentiation (LTP) is a key mechanism underlying neural plasticity and learning. This phenomenon occurs when repeated stimulation of a synapse enhances its strength, making it more responsive to future signals. LTP is believed to be the cellular basis of memory and learning, as it enables the brain to retain information over time. Conversely, long-term depression (LTD) involves the weakening of synaptic connections, allowing the brain to discard outdated or irrelevant information. Together, these processes enable the brain to strike a balance between retaining valuable knowledge and maintaining flexibility for new learning.

The brain's capacity for plasticity is not limited to the strengthening and weakening of synapses; it also extends to the formation of new neurons, a process known as neurogenesis. While neurogenesis occurs primarily in specific regions of the brain, such as the hippocampus, it contributes to learning and memory by providing new cells that can integrate into existing neural circuits. Factors such as physical exercise, enriched environments, and certain dietary components have been shown to promote neurogenesis, highlighting the interplay between lifestyle and brain health.

The role of neural plasticity in learning becomes particularly evident in the context of skill acquisition and practice. When learning a new skill, such as playing a musical instrument or mastering a sport, the brain undergoes structural and functional changes to accommodate the demands of the activity. Repeated practice strengthens the neural circuits involved, leading to greater precision and efficiency. Over time, these changes can become so ingrained that the skill becomes automatic, requiring minimal conscious effort. This phenomenon, known as procedural learning, demonstrates the brain's ability to optimize its resources through plasticity.

Neural plasticity also plays a crucial role in recovery from injury or trauma. When a part of the brain is damaged, such as after a stroke or traumatic brain injury, other regions can compensate for the loss by reorganizing neural circuits. This process, known as functional plasticity, allows individuals to regain abilities that were thought to be permanently lost. Rehabilitation and targeted therapies can enhance this recovery by stimulating neural pathways and encouraging the formation of new connections. The brain's resilience in the face of injury underscores the profound adaptability afforded by neural plasticity.

Environmental factors and lifestyle choices can significantly influence the extent and effectiveness of neural plasticity. Engaging in mentally stimulating activities, such as reading, problem-solving, or learning a new language, can enhance plasticity by challenging the brain and promoting the formation of new connections. Physical exercise has been shown to improve neural plasticity by increasing blood flow to the brain and stimulating the release of growth factors that support neuronal health. Adequate sleep is another critical factor, as it provides the brain with the opportunity to consolidate memories and repair neural networks. Conversely, chronic stress, poor nutrition, and a sedentary lifestyle can impair neural plasticity, hindering the brain's ability to adapt and learn.

While neural plasticity offers immense potential for growth and adaptation, it is not without its limitations. Certain factors, such as aging and neurodegenerative diseases, can reduce the brain's plasticity over time, making learning and recovery more challenging. However, research continues to uncover strategies for enhancing plasticity, such as cognitive training, pharmacological interventions, and lifestyle modifications. These advancements hold promise for mitigating the effects of aging and disease, as well as unlocking the brain's full potential for lifelong learning.

The interplay between neural plasticity and learning reveals the profound adaptability of the human brain.

By forming, strengthening, and refining connections, the brain continually evolves in response to new experiences and challenges. This dynamic process underscores the importance of engaging in activities that promote brain health and stimulate neural pathways, ensuring that the capacity for learning and adaptation remains robust throughout life. Neural plasticity is a testament to the brain's resilience and ingenuity, a mechanism that allows individuals to grow, change, and thrive in an ever-changing world.

Chapter 5: Executive Functions and Higher-Order Thinking

Problem Solving Networks

Problem-solving relies on the intricate networks of the brain working together to analyze, process, and resolve challenges. These networks are not isolated entities; they represent an interconnected system of regions that collaborate to identify problems, generate potential solutions, and execute decisions. The brain, as a complex organ, utilizes a variety of mechanisms and pathways to address both routine tasks and novel issues, drawing from past experiences, logical reasoning, creativity, and adaptability. Understanding these networks sheds light on how humans navigate challenges, innovate, and adapt to an ever-changing world.

At the core of problem-solving lies the interaction between two primary brain networks: the default mode network (DMN) and the executive control network (ECN). The DMN is associated with introspection, daydreaming, and generating ideas. It is most active when the brain is at rest or engaged in spontaneous thought processes, often playing a significant role in creative problem-solving. The ECN, on the other hand, is responsible for focused attention, planning, and decision-making. It becomes active during tasks requiring concentration and analytical reasoning. The interplay between these networks allows the brain to oscillate between divergent thinking, where multiple possible solutions are generated, and convergent thinking, where the most appropriate solution is selected and implemented.

The salience network acts as a bridge between the DMN and ECN, enabling the brain to switch between inward-focused and outward-focused modes of thought. This network evaluates incoming stimuli to determine their relevance and importance, ensuring the brain allocates resources to address the problem at hand. For instance, when faced with a sudden and urgent issue, the salience network helps direct attention to the immediate challenge, activating the ECN to focus on logical and practical solutions. Conversely, when the situation requires creativity or long-term planning, the salience network allows for a shift toward the DMN, fostering innovative approaches.

Memory and learning also play vital roles in problem-solving, as they provide the foundation upon which solutions are built. The hippocampus, a region heavily involved in memory, retrieves relevant information from past experiences to inform decision-making. By recalling similar situations or drawing upon knowledge previously acquired, the brain can apply known strategies and adapt them to new challenges. This reliance on memory highlights the importance of experience in honing problem-solving abilities; individuals who have encountered a wide variety of situations are often more adept at identifying effective solutions.

Emotion is another critical factor that influences problem-solving networks, often acting as either a catalyst or a barrier. The amygdala, which processes emotions, interacts with the prefrontal cortex to regulate emotional responses and assess the stakes of a given problem. In certain scenarios, emotions such as fear or urgency can heighten focus and drive the brain to resolve an issue more quickly. However, overwhelming emotions, such as anxiety or frustration, can impair cognitive functions, making it more difficult to think clearly or assess options rationally. The brain's ability to manage and regulate emotions is therefore essential for effective problem-solving.

Creativity emerges as a powerful tool within problem-solving, particularly when conventional methods prove insufficient. The brain's ability to form new connections and think outside established patterns often stems from the interaction of diverse networks working in tandem. The anterior cingulate cortex, for instance, plays a pivotal role in detecting conflicts or inconsistencies, prompting the brain to consider alternative approaches. Simultaneously, the temporal lobes contribute to pattern recognition, while the prefrontal cortex evaluates the feasibility of novel ideas. This collaboration fosters innovation, allowing the brain to generate unique solutions that might not have been immediately apparent.

Problem-solving is not solely a cognitive process; it is also influenced by the body and environment. Physical movement, such as walking or engaging in light exercise, has been shown to enhance the brain's ability to generate ideas and solve problems. The increased blood flow and release of endorphins during physical activity can improve focus and creativity, while the change in surroundings can provide new perspectives. Similarly, external factors such as social interaction and collaboration can enhance problem-solving by introducing diverse viewpoints and pooling collective knowledge.

The brain's problem-solving networks also demonstrate remarkable adaptability, allowing individuals to respond to challenges in real-time. This adaptability is evident in the brain's ability to reconfigure itself in response to novel or complex problems. When faced with an unfamiliar situation, the brain often recruits additional regions to support the effort, creating temporary networks that facilitate the resolution of the issue. This flexibility underscores the dynamic nature of problem-solving, as the brain continuously adjusts its strategies to meet the demands of each unique challenge.

However, problem-solving networks are not infallible. Cognitive biases, such as confirmation bias or anchoring, can distort perception and lead to flawed decisions. These biases often arise from the brain's tendency to rely on heuristics, or mental shortcuts, to simplify complex problems. While heuristics can be useful in certain contexts, they can also result in errors or oversights. Awareness of these biases and a deliberate effort to challenge assumptions can help mitigate their impact, enabling more accurate and effective problem-solving.

The development and refinement of problem-solving skills are closely tied to the brain's plasticity, as repeated practice and exposure to challenges strengthen the underlying networks. Activities such as puzzles, strategic games, and critical thinking exercises can enhance the brain's capacity to analyze and resolve problems. Additionally, fostering an open and curious mindset encourages exploration and experimentation, which are essential for creative and adaptive problem-solving.

The intricate collaboration of problem-solving networks illustrates the brain's remarkable capacity to address challenges, innovate, and adapt. By leveraging memory, emotion, creativity, and adaptability, the brain navigates the complexities of decision-making and resolution. These networks represent not only the essence of human cognition but also the foundation of progress and growth, enabling individuals to overcome obstacles and seize opportunities in an ever-changing world.

Critical Thinking Pathways

Critical thinking is a multifaceted process that allows individuals to assess information, analyze situations, and form well-reasoned judgments. It is not merely a skill but a framework of cognitive pathways that work together to evaluate facts, challenge assumptions, and draw conclusions. The ability to think critically is crucial for navigating the complexities of modern life, where information is abundant but often contradictory or misleading. By engaging specific neural networks, the brain enables us to question, reason, and arrive at thoughtful decisions.

At the heart of critical thinking lies the prefrontal cortex, a region of the brain responsible for higher-order cognitive functions such as decision-making, planning, and self-reflection. This area is particularly active when we process information that requires careful evaluation or complex reasoning. The brain's ability to weigh evidence, identify patterns, and anticipate outcomes is rooted in the flexibility of neural connections within this region. When faced with a problem or question, the prefrontal cortex collaborates with other areas of the brain to gather and analyze relevant data, ensuring that conclusions are based on logic rather than impulse.

Memory plays a foundational role in critical thinking by providing the context and knowledge necessary to evaluate new information. The hippocampus retrieves relevant experiences, facts, and learned concepts, allowing the brain to compare current situations with past ones. This comparison is essential for recognizing patterns, identifying inconsistencies, and making informed judgments. For example, when assessing the credibility of a source, the brain draws upon prior knowledge about reliable indicators of trustworthiness, such as expertise, evidence, and consistency. Without this ability to reference past experiences, critical thinking would lack the depth and precision needed for effective decision-making.

Emotional regulation also influences critical thinking pathways, as emotions can either enhance or hinder the process. The interaction between the amygdala and the prefrontal cortex allows the brain to assess emotional responses and integrate them into rational thought. While emotions like fear or anger can cloud judgment and lead to biased decisions, the ability to recognize and manage these feelings is a key component of critical thinking. By maintaining emotional balance, individuals can approach problems with a clear and focused mindset, minimizing the impact of cognitive distortions.

One of the most important aspects of critical thinking is the ability to recognize and challenge assumptions. Assumptions often operate below the level of conscious awareness, shaping perceptions and influencing decisions without scrutiny. The brain's anterior cingulate cortex plays a significant role in detecting conflicts or discrepancies between expectations and reality. When an assumption is questioned, this region activates and prompts the brain to reconsider its initial perspective. For instance, encountering evidence that contradicts a deeply held belief can trigger a reassessment of that belief, fostering a more nuanced and accurate understanding.

The evaluation of evidence is another cornerstone of critical thinking, requiring the brain to distinguish between relevant and irrelevant information. The parietal lobes contribute to this process by processing sensory input and categorizing data based on its significance. Meanwhile, the prefrontal cortex assesses the credibility and reliability of information sources. This dual process ensures that conclusions are grounded in accurate and pertinent evidence. For example, when analyzing a scientific study, critical thinkers evaluate the methodology, sample size, and potential biases to determine the validity of the findings.

Critical thinking also involves the ability to consider multiple perspectives and alternative solutions. This cognitive flexibility is supported by the brain's ability to form and reorganize neural connections, a phenomenon known as neuroplasticity. By exploring different viewpoints, the brain broadens its understanding of a problem and increases the likelihood of finding innovative solutions. The temporoparietal junction, a region associated with perspective-taking and empathy, plays a key role in this process by allowing individuals to consider how others might perceive a situation. This capacity to step outside one's own perspective enhances problem-solving and decision-making, particularly in collaborative or interpersonal contexts.

Biases and cognitive shortcuts can pose significant challenges to critical thinking, as they often lead to flawed or incomplete conclusions. The brain's reliance on heuristics, or mental shortcuts, is a natural response to the need for efficiency in processing vast amounts of information. However, these shortcuts can result in errors such as confirmation bias, where individuals favor information that supports their preexisting beliefs, or anchoring bias, where initial impressions disproportionately influence subsequent judgments. Recognizing and counteracting these biases requires deliberate effort and self-awareness. The brain's ability to monitor and adjust its own thought processes, known as metacognition, is essential for overcoming these obstacles and achieving more accurate and balanced conclusions.

The role of creativity in critical thinking is often underestimated but is equally essential. Creativity allows the brain to generate novel ideas and approaches, particularly when conventional methods fall short. The interaction between the default mode network, associated with spontaneous thought, and the executive control network, responsible for focused reasoning, fosters this blend of creativity and logic. For instance, solving a complex problem may require not only analytical skills but also the ability to think outside the box and consider unconventional solutions. This interplay between structured and imaginative thinking enables the brain to navigate ambiguity and uncertainty effectively.

Developing critical thinking pathways is a lifelong process that requires practice, reflection, and a commitment to intellectual growth. Engaging in activities that challenge the brain, such as reading, debating, or solving puzzles, strengthens the neural networks involved in reasoning and analysis. Exposure to diverse perspectives and experiences also enhances the brain's ability to evaluate information and adapt to new challenges. By cultivating habits of curiosity, skepticism, and open-mindedness, individuals can refine their critical thinking skills and apply them to a wide range of contexts.

The dynamic interplay of memory, emotion, logic, and creativity within critical thinking pathways underscores the brain's remarkable capacity to navigate complexity. These pathways enable individuals to question assumptions, evaluate evidence, and consider alternative perspectives, fostering a deeper understanding of the world and more effective decision-making. By honing these cognitive processes, individuals can approach challenges with clarity, adaptability, and confidence, ensuring that their judgments are as thoughtful and informed as possible.

Creativity and Brain Integration

Creativity stems from the intricate interplay of various regions of the brain, working together in a delicate yet dynamic balance. Contrary to the old notion of a strict left-brain versus right-brain dichotomy, modern neuroscience reveals that creativity is the result of integration across multiple neural networks. This collaboration between different regions allows the mind to generate novel ideas, solve problems in innovative ways, and express itself in ways that transcend conventional thinking. Understanding how the brain integrates its components to foster creativity not only provides insights into human potential but also offers practical strategies to enhance creative capabilities.

At the heart of creativity lies the default mode network (DMN), often referred to as the brain's "imagination network." This network becomes active during periods of rest, daydreaming, and introspection, allowing the mind to wander into unexplored territories. The DMN connects various brain regions, including the medial prefrontal cortex, posterior cingulate cortex, and angular gyrus. By fostering connections between seemingly unrelated ideas or memories, this network lays the groundwork for creative thinking. However, the DMN does not act alone. It must interact with other networks, such as the executive control network (ECN) and the salience network, to refine and implement creative ideas.

The executive control network plays a crucial role in focusing attention, evaluating ideas, and ensuring that creative thoughts are practical or relevant. Without the ECN's regulatory function, the raw, unfiltered output of the imagination network might remain chaotic and unproductive. This network involves the prefrontal cortex and other regions that regulate higher-order cognitive functions, such as planning and decision-making. When someone is engaged in a creative task, the ECN steps in to assess possibilities and determine which ideas have the most potential. This delicate balance between free-flowing imagination and structured analysis is one of the cornerstones of creativity.

The salience network acts as a bridge between the DMN and the ECN, determining what stimuli or ideas are worth focusing on. Comprising the anterior cingulate cortex and the insula, this network identifies patterns, detects changes in the environment, and prioritizes information. For example, when a person encounters an intriguing concept or an inspiring image, the salience network flags it as significant, directing attention toward it. By filtering out distractions and highlighting relevant inputs, this network ensures that creative energy is channeled effectively.

Another essential aspect of brain integration in creativity is the role of neuroplasticity, the brain's ability to adapt and form new connections. Engaging in diverse experiences, learning new skills, and exposing oneself to novel environments stimulate neuroplasticity, creating pathways that facilitate creative thinking. When the brain encounters fresh stimuli, it forges connections between previously unrelated ideas, leading to innovative insights. This adaptability underscores the importance of curiosity and exploration in nurturing creativity. A mind that remains open to new possibilities is better equipped to integrate disparate elements into cohesive, original concepts.

Emotions also play a significant role in shaping creativity. The limbic system, which governs emotions, interacts with cognitive networks to influence creative output. Positive emotions, such as joy and excitement, enhance the brain's capacity to associate ideas and expand its scope of thinking. Negative emotions, while often seen as hindrances, can also fuel creativity by providing depth and perspective. For instance, periods of introspection during challenging times can lead to profound artistic or intellectual breakthroughs. The integration of emotional and cognitive processes enriches creative expression, adding depth and authenticity to ideas.

Another factor that supports creativity is the synchronization of brain hemispheres. While it is true that certain functions, such as language and spatial awareness, are more dominant in one hemisphere, creativity often requires collaboration between both sides. The corpus callosum, a bundle of nerve fibers connecting the two hemispheres, plays a pivotal role in this process. By facilitating communication between the analytical left brain and the holistic right brain, the corpus callosum enables the synthesis of logical reasoning with imaginative thinking. This interplay is particularly evident in activities such as music, where rhythm and melody must coexist in harmony, or in problem-solving, where both linear and lateral thinking are necessary.

Mindfulness and relaxation techniques can further enhance brain integration, creating an optimal state for creativity to flourish. Practices such as meditation, deep breathing, and even physical exercise reduce stress and promote the synchronization of neural networks. When the mind is calm and focused, it is better able to access the resources of the imagination network while maintaining the discipline of the executive control network. This state of "flow," where the mind is fully immersed in a creative task, is often described as one of the most fulfilling human experiences. During flow, the boundaries between thought and action blur, and ideas seem to emerge effortlessly, as if the brain's various components are working together in perfect harmony.

Sleep is another critical factor in fostering creativity and brain integration. During sleep, particularly the rapid eye movement (REM) stage, the brain consolidates memories, processes emotions, and forms new connections. Many creative breakthroughs have been attributed to insights gained during or after sleep, when the mind has had the opportunity to organize and integrate information. Dreams, a product of the imagination network in action, often serve as a wellspring of inspiration, offering new perspectives on familiar problems.

Fostering creativity through brain integration requires not only an understanding of these neural mechanisms but also the cultivation of habits that support them. Engaging in activities that challenge the mind, such as solving puzzles, learning a new language, or experimenting with different art forms, keeps the brain flexible and responsive. Building a lifestyle that includes periods of reflection, exploration, and focused effort creates the conditions for creativity to thrive. By appreciating the complex interplay of neural networks and embracing practices that nurture their integration, individuals can unlock their creative potential and bring their most innovative ideas to light.

Executive Control Systems

The executive control systems of the brain act as the command center for human behavior, guiding actions, thoughts, and decisions with precision and adaptability. These systems, often associated with the prefrontal cortex, represent the pinnacle of cognitive function, enabling individuals to navigate complex environments, achieve goals, and regulate impulses. They are not isolated mechanisms but rather a network of interconnected processes that work in harmony to ensure the efficient management of information and responses. Understanding the intricacies of these systems sheds light on how humans plan, adapt, and maintain focus amidst an ever-changing world.

At the core of executive control lies the ability to prioritize and manage tasks. This process involves working memory, which allows for the temporary storage and manipulation of information. Working memory operates like a mental workspace, holding relevant details while filtering out distractions. For instance, when solving a problem or making a decision, the mind relies on this capacity to organize data, compare options, and foresee potential outcomes. The dorsolateral prefrontal cortex plays a central role in this function, acting as a hub for integrating and processing information in real-time. Without working memory, even the simplest tasks would become overwhelming, as the brain would struggle to retain and apply relevant knowledge.

Flexibility is another hallmark of executive control systems. The ability to shift perspectives, adapt strategies, and pivot between tasks is essential in a world that demands constant adjustment. Cognitive flexibility, as it is often called, enables individuals to approach challenges from multiple angles and embrace new solutions when old ones fail. This adaptability is rooted in the brain's capacity to reconfigure neural connections, allowing for the seamless transition between modes of thought. The anterior cingulate cortex, a key component of the executive control network, monitors for conflicts and errors, signaling the need for adjustments. Whether it's switching between professional and personal priorities or finding new ways to solve a problem, cognitive flexibility ensures that the mind remains agile and responsive.

Self-regulation is another critical function of executive control systems. The ability to manage emotions, suppress impulses, and delay gratification is essential for achieving long-term goals. This process involves the ventromedial prefrontal cortex, which interacts with the amygdala and other emotional centers to maintain balance. For example, when confronted with a tempting distraction, such as a notification on a phone during an important task, the executive control systems intervene to suppress the immediate urge and redirect attention to the priority at hand. This capacity for self-regulation not only supports productivity but also fosters resilience and emotional stability.

Decision-making is another domain where executive control systems shine. Decisions often require the evaluation of multiple variables, weighing risks and rewards, and predicting future consequences. The orbitofrontal cortex plays a pivotal role in this process, integrating sensory input, memories, and emotional cues to guide choices. For instance, when deciding whether to invest time in a new project or stick with an existing one, the brain assesses potential benefits and drawbacks, drawing on past experiences and current circumstances. This intricate calculus highlights the sophistication of executive control systems, which balance logic and intuition to arrive at optimal outcomes.

Attention control is fundamental to the effectiveness of executive systems. The ability to focus on relevant stimuli while ignoring distractions is a skill honed by the brain's attentional networks. The prefrontal cortex collaborates with the parietal lobe to direct focus, ensuring that cognitive resources are allocated to the task at hand. In an age where constant notifications and information overload compete for attention, the importance of this function cannot be overstated.

Whether it's maintaining concentration during a presentation or staying immersed in creative work, attention control underscores the brain's capacity to filter and prioritize information.

Motivation, too, is deeply intertwined with executive control. The drive to pursue goals, overcome obstacles, and persist in the face of challenges is fueled by the interaction between the prefrontal cortex and the brain's reward systems. Dopamine, a neurotransmitter associated with pleasure and reward, plays a significant role in sustaining motivation, reinforcing behaviors that lead to success. For example, the satisfaction of completing a challenging project or achieving a personal milestone activates these reward circuits, encouraging further effort. By linking effort with reward, the executive systems maintain momentum and ensure progress.

The development and refinement of executive control systems are influenced by various factors, including age, experience, and environment. During childhood and adolescence, these systems undergo significant growth and maturation, as neural connections are strengthened and refined. Experiences that challenge the brain, such as learning new skills or solving complex problems, further enhance executive function by fostering adaptability and resilience. Conversely, stress, fatigue, and other adverse conditions can impair these systems, highlighting the importance of maintaining a supportive environment for optimal performance.

While the executive control systems are remarkably sophisticated, they are not infallible. Cognitive biases, emotional interference, and external pressures can undermine their effectiveness, leading to errors in judgment and lapses in self-control. Recognizing these limitations is crucial for developing strategies to mitigate their impact. Practices such as mindfulness, regular physical activity, and adequate sleep have been shown to bolster executive function, enhancing focus, decision-making, and emotional regulation. By nurturing these systems, individuals can optimize their cognitive capabilities and achieve their full potential.

The interplay of neural networks within the executive control systems underscores the complexity and adaptability of the human mind. These systems, far from operating in isolation, reflect the brain's ability to coordinate and integrate diverse functions to navigate the demands of everyday life. Whether it's managing competing priorities, solving intricate problems, or pursuing long-term aspirations, the executive control systems serve as the foundation for purposeful and directed behavior. Their capacity to adapt, regulate, and innovate is a testament to the remarkable resilience and ingenuity of the human brain.

Metacognition Processes

Metacognition, often described as "thinking about thinking," is the process that allows individuals to reflect on, monitor, and regulate their cognitive activities. It encompasses a deep awareness of one's own thought processes, enabling the evaluation of how knowledge is acquired, stored, and applied. This self-regulatory mechanism is fundamental to learning, problem-solving, and decision-making, offering a framework for understanding not only what we know but also how we know it. By engaging in metacognitive processes, individuals can adapt their strategies, improve their understanding, and achieve greater control over their mental faculties.

At its core, metacognition is split into two interrelated components: metacognitive knowledge and metacognitive regulation. Metacognitive knowledge refers to the awareness of one's cognitive abilities, the nature of tasks, and the strategies available to approach those tasks. For example, understanding that certain types of problems require logical reasoning while others demand creative thinking is a form of metacognitive knowledge. This awareness extends to recognizing personal strengths and weaknesses, such as knowing whether one tends to procrastinate or struggles with focus. On the other hand, metacognitive regulation involves actively overseeing and adjusting one's cognitive processes in real time. This includes planning how to approach a task, monitoring progress, and making adjustments when encountering obstacles. Together, these components form a dynamic process that enhances cognitive performance.

The importance of metacognition becomes evident in the context of learning. Students who possess strong metacognitive skills are often better equipped to set realistic goals, choose effective study strategies, and assess their understanding of material. For instance, a student preparing for an exam might evaluate their comprehension of a subject, identify areas of weakness, and decide to allocate more time to those topics. During the study session, they might periodically pause to test their recall or seek clarification on concepts they find confusing. This continuous cycle of planning, monitoring, and evaluating ensures a more efficient and targeted approach to learning, reducing the likelihood of wasted effort.

Metacognition also plays a pivotal role in problem-solving. When faced with a challenge, individuals with well-developed metacognitive abilities can assess the complexity of the task, determine the resources needed, and devise a strategy to tackle it. They remain aware of their progress, identifying when a particular approach is ineffective and pivoting to alternative methods. This adaptability is especially valuable in situations where the solution is not immediately apparent or requires creative thinking. By reflecting on past experiences and applying lessons learned, metacognitive thinkers can navigate uncertainty with greater confidence and precision.

One of the most intriguing aspects of metacognition is its role in fostering self-awareness. The ability to step back and observe one's thoughts, emotions, and behaviors creates a space for introspection and growth. This self-awareness extends beyond cognitive processes, encompassing emotional regulation and interpersonal interactions. For example, someone engaged in a heated argument might recognize their escalating frustration and choose to pause and recalibrate their response. This moment of reflection, driven by metacognitive processes, can prevent impulsive reactions and promote more constructive communication. In this way, metacognition serves as a bridge between thought and action, allowing individuals to align their behavior with their intentions.

While metacognition is a powerful tool for enhancing cognitive and emotional functioning, it is not an innate ability that comes effortlessly to everyone. Like any skill, it requires practice and cultivation. Developing metacognitive awareness begins with fostering curiosity and a willingness to question one's assumptions. Asking reflective questions such as "What am I trying to achieve?" or "Is this approach working?" can initiate the process of self-monitoring and regulation. Over time, these practices become habitual, ingraining metacognitive processes into daily life.

The benefits of metacognition extend beyond individual contexts, influencing how people interact with the world around them. In collaborative settings, metacognitive thinkers can enhance group dynamics by recognizing the strengths and weaknesses of team members, facilitating effective communication, and ensuring that collective goals are met. For example, in a workplace project, someone with strong metacognitive skills might notice when the team is veering off track and propose a course correction. By reflecting on group processes and encouraging others to do the same, they contribute to a more cohesive and productive environment.

Metacognition also intersects with creativity, as it allows individuals to evaluate the originality and feasibility of their ideas. Creative endeavors often involve navigating ambiguity and exploring uncharted territory, which can be both exhilarating and daunting. Metacognitive processes enable creators to assess their progress, identify when they are stuck in unproductive patterns, and seek inspiration from new sources. This iterative process of reflection and adjustment fuels innovation, ensuring that creative efforts remain purposeful and impactful.

Despite its numerous advantages, metacognition is not without its challenges. Overthinking or excessive self-monitoring can lead to paralysis by analysis, where individuals become so preoccupied with evaluating their thoughts that they struggle to take action. Striking a balance between reflection and execution is essential to harness the full potential of metacognition. Additionally, cultural and educational factors can influence the development of metacognitive skills, highlighting the need for intentional teaching and support. Encouraging metacognitive practices in schools, workplaces, and communities can empower individuals to think critically and adaptively, fostering resilience and lifelong learning.

Chapter 6: Development and Aging of the Mind

Neural Development Patterns

Neural development is a complex and transformative process that begins long before birth and continues throughout life. It shapes the architecture of the brain, laying the foundation for cognition, emotion, and behavior. Patterns of neural development are influenced by a combination of genetic instructions and environmental experiences, creating a dynamic interplay that determines the structure and function of the nervous system. Understanding these patterns not only reveals the intricate design of the human brain but also highlights the factors that contribute to its remarkable adaptability and potential for change.

The early stages of neural development begin during embryogenesis, when the neural tube forms as one of the first structures of the central nervous system. This tube eventually gives rise to the brain and spinal cord. Cellular differentiation occurs as stem cells specialize into neurons and glial cells, the primary building blocks of the nervous system. During this phase, neurons are generated in large numbers, a process known as neurogenesis, which peaks during fetal development. These neurons migrate to their designated locations, guided by chemical signals and the physical scaffolding provided by radial glial cells. This precise orchestration ensures that different regions of the brain are populated by the appropriate types of neurons needed for their specific functions.

As neurons settle into their target regions, they begin forming connections with one another through axons and dendrites. These connections, known as synapses, are the communication points between neurons and form the foundation of neural circuits. Synaptogenesis, or the formation of synapses, occurs at an extraordinary rate during early development, with an infant's brain forming millions of new connections each second. This explosion of synaptic growth is accompanied by a period of synaptic pruning, during which weaker or unused connections are eliminated. Synaptic pruning is critical for optimizing brain efficiency, as it refines neural networks to strengthen the most frequently used pathways while discarding redundant or unnecessary ones.

Myelination is another key pattern of neural development, involving the formation of a fatty sheath around the axons of neurons. This sheath, known as myelin, greatly enhances the speed and efficiency of electrical signals traveling along neural pathways. Myelination begins during the prenatal stage and continues well into adulthood, with different regions of the brain myelinating at different rates. For example, the sensory and motor areas of the brain undergo myelination earlier than the prefrontal cortex, which governs higher-order functions such as reasoning and decision-making. This staggered timeline reflects the brain's prioritization of basic survival skills in infancy before progressing to more complex cognitive abilities in later years.

The brain's plasticity, or its ability to adapt and reorganize itself, is a defining feature of neural development. During critical periods of development, the brain is particularly sensitive to environmental stimuli, which shape neural circuits and influence long-term outcomes. For instance, the development of vision depends on the proper stimulation of the visual cortex during infancy. If an infant's visual experience is disrupted during this critical period, such as by untreated cataracts, the brain's ability to process visual information may be permanently impaired. Similarly, language acquisition relies on exposure to spoken language during early childhood, when the brain is most receptive to forming the neural connections necessary for communication. These critical periods underscore the importance of timely and enriching experiences in shaping the brain's developmental trajectory.

Environmental factors play a significant role in neural development, interacting with genetic predispositions to influence outcomes. Nutrition is one such factor, as the brain requires specific nutrients, such as omega-3 fatty acids and iron, to support growth and function. During pregnancy and early childhood, malnutrition can hinder neurogenesis, myelination, and synaptic formation, leading to long-term cognitive and behavioral deficits. Similarly, exposure to toxins, such as lead or alcohol, can disrupt neural development by interfering with cellular processes or damaging existing connections. On the other hand, positive environmental factors, such as responsive caregiving, stimulating activities, and access to education, can enhance neural development by promoting the formation and strengthening of beneficial neural pathways.

Social interactions also play a profound role in shaping neural development. The brain is inherently social, with certain regions, such as the prefrontal cortex and the amygdala, specialized for processing social and emotional information. Early relationships with caregivers provide the foundation for social and emotional learning, influencing the development of attachment, empathy, and self-regulation. For instance, a child who experiences consistent and nurturing interactions is more likely to develop secure attachments and adaptive stress responses, supported by well-integrated neural circuits. Conversely, adverse experiences, such as neglect or abuse, can disrupt the development of these circuits, increasing the risk of emotional and behavioral difficulties.

During adolescence, neural development enters a period of significant reorganization, driven by hormonal changes and continued synaptic pruning. The prefrontal cortex, responsible for executive functions such as planning and impulse control, undergoes substantial maturation during this time. However, its development lags behind that of the limbic system, which governs emotions and reward processing. This imbalance explains the heightened risk-taking and emotional volatility often observed during adolescence, as the brain's regulatory systems have not yet fully matured. Despite these challenges, adolescence is also a time of heightened plasticity, offering opportunities for learning, creativity, and personal growth.

Neural development does not cease with adulthood but continues throughout life in response to experiences and challenges. While the pace of neurogenesis and plasticity slows with age, the brain retains the capacity to form new connections, adapt to changes, and recover from injury. Lifelong learning, physical activity, and social engagement have been shown to promote brain health by stimulating neuroplasticity and reducing the risk of cognitive decline. This ongoing adaptability highlights the resilience of the human brain and its ability to evolve in response to the demands of a changing environment.

The patterns of neural development, from the rapid growth of infancy to the refinement of adulthood, reflect the brain's extraordinary capacity for adaptation and learning. These intricate processes are shaped by a combination of genetic blueprints and environmental influences, working together to create the complex networks that underlie human thought, emotion, and behavior. By understanding these patterns, we gain insight into the factors that promote healthy development and the interventions that can support individuals at every stage of life. Neural development is not merely a biological phenomenon; it is a testament to the profound interconnectedness of nature and nurture in shaping the human experience.

Cognitive Milestones

Cognitive milestones represent significant markers in the development of thought processes, problem-solving abilities, memory, and language comprehension. These benchmarks, while often generalized by age, are not rigidly fixed, as individual differences and environmental factors play a crucial role in shaping the pace and nature of cognitive growth. From infancy to adulthood, the progression of these milestones provides insight into how humans acquire, organize, and apply knowledge, highlighting the dynamic interplay between biological maturation and experiential learning.

The journey begins in infancy, a period marked by rapid and profound changes in cognitive abilities. Newborns arrive equipped with basic reflexes and a limited capacity for sensory processing, but their brains are primed for exploration and learning. Within the first few months, infants begin to demonstrate recognition of familiar faces and voices, a milestone that signifies early memory formation. They also engage in behaviors such as tracking moving objects with their eyes and responding to auditory stimuli, laying the foundation for visual and auditory processing. By six months of age, a significant milestone emerges as infants develop object permanence, the understanding that objects continue to exist even when they are out of sight. This concept, though seemingly simple, reflects an important leap in memory and reasoning, as it requires the infant to hold a mental representation of the unseen object.

The toddler years bring a surge in cognitive abilities, fueled by the rapid expansion of vocabulary and the emergence of symbolic thinking. By their second year, most toddlers begin to use words to represent objects, actions, and ideas, a development that transforms the way they interact with their environment. The ability to engage in pretend play, such as using a block to represent a car or a doll to mimic caregiving, indicates the growing complexity of their mental representations. Problem-solving skills also take a significant step forward during this stage, as toddlers become adept at trial-and-error learning. For example, they may experiment with stacking blocks in different configurations to create a stable tower, demonstrating an understanding of cause-and-effect relationships.

Preschool-aged children continue to build on these foundational skills, entering what developmental psychologist Jean Piaget termed the preoperational stage. During this phase, children display remarkable advances in language and imagination but remain limited by egocentric thinking, which makes it challenging for them to see things from perspectives other than their own. They engage in increasingly complex pretend play, constructing elaborate narratives and scenarios that reflect their expanding cognitive and social understanding. Another milestone of this stage is the ability to classify objects based on shared characteristics, such as sorting shapes by color or size. While their reasoning is still influenced by concrete and immediate experiences, preschoolers begin to grasp concepts such as quantity and time, even if their understanding remains rudimentary.

The transition to middle childhood brings a shift toward more logical and organized thinking, as children enter the concrete operational stage. At this point, they achieve milestones such as the ability to perform mental operations, including addition and subtraction, and to understand the principles of conservation. Conservation refers to the realization that certain properties of objects, such as volume or mass, remain constant despite changes in appearance. For instance, a child who understands conservation will recognize that water poured from a wide cup into a tall, narrow glass remains the same amount, even though it looks different. This stage also marks the development of classification and seriation skills, enabling children to organize objects into hierarchical groups or arrange them in a specific order based on measurable attributes.

As children approach adolescence, their cognitive abilities continue to mature, culminating in the capacity for abstract and hypothetical thinking. This stage, known as formal operational thinking, allows adolescents to consider multiple possibilities, analyze complex relationships, and engage in deductive reasoning. They begin to explore abstract concepts such as justice, morality, and identity, often grappling with questions that have no definitive answers. This newfound cognitive flexibility supports critical thinking and problem-solving, as adolescents become capable of evaluating evidence, constructing arguments, and testing hypotheses. While these abilities signify a significant milestone in intellectual development, they are accompanied by an increased awareness of self and others, which can sometimes lead to heightened self-consciousness or susceptibility to peer influence.

Adulthood introduces its own set of cognitive milestones, many of which are shaped by the demands of personal and professional life. Young adults often reach the peak of their problem-solving abilities, drawing on both logical reasoning and experiential knowledge to navigate challenges. They become adept at integrating information from multiple sources, a skill that is particularly valuable in decision-making and creative endeavors. As individuals progress through adulthood, their cognitive priorities may shift, with a greater emphasis on wisdom and practical intelligence. These milestones reflect the accumulation of life experience and the ability to apply knowledge in meaningful and adaptive ways.

While cognitive development is often associated with growth and achievement, it is important to recognize that aging can bring changes to cognitive function. Some abilities, such as processing speed and working memory, may decline over time, but others, such as vocabulary and general knowledge, often remain stable or even improve. The concept of cognitive reserve highlights the brain's capacity to adapt and compensate for age-related changes, suggesting that lifelong learning, social engagement, and physical activity can help preserve cognitive function in later years.

Cognitive milestones are not merely markers of development but are also windows into the intricate workings of the human mind. They illustrate the ways in which individuals interact with their environments, construct meaning, and solve problems, highlighting the adaptability and resilience of the brain. While these milestones follow general patterns, they are deeply influenced by the unique experiences and opportunities that shape each person's journey. By understanding and supporting the progression of these milestones, we can foster environments that

nurture curiosity, learning, and growth across all
stages of life.

Brain Maturation Process

The process of brain maturation is a marvel of
biological precision, a journey that spans from the
earliest stages of fetal development to the later years
of adulthood. It is a dynamic interplay of genetic
programming and environmental influences, where
the brain undergoes structural and functional changes
that shape how individuals think, feel, and behave.
Each stage builds upon the foundations laid before it,
with distinct milestones marking the progression
toward a fully mature and integrated brain. This
process is far from linear; it is highly adaptive,
responding to the demands of the environment and
the experiences one encounters over time.

Brain maturation begins in utero, with the formation
of the neural tube during the first few weeks of
pregnancy, which later develops into the brain and
spinal cord. Rapid cellular proliferation follows, as
neurons are generated in excess, far exceeding the
number that will ultimately remain in the adult brain.
These neurons migrate to specific regions, guided by
genetic cues and chemical signals, to form the basic
architecture of the brain. By the time of birth, the
brain has established its foundational structure, but it
is still far from fully developed. While the number of
neurons has largely stabilized, the connections
between them—synapses—are only just beginning to
form in earnest.

The first years of life are characterized by an explosion of synaptic connections, a process known as synaptogenesis. During this period, the brain is incredibly plastic, meaning it has a heightened capacity for forming and reorganizing neural pathways in response to stimuli. Infants are born with reflexive behaviors, but as they interact with their environment, their brains quickly adapt, forming connections that support sensory processing, motor skills, and early cognitive functions. The sensory and motor cortices are among the first areas to mature, enabling infants to coordinate movements and respond to visual, auditory, and tactile stimuli. This early period is critical for establishing the neural groundwork for more complex functions.

As children grow, their brains undergo a process of refinement. This is achieved primarily through synaptic pruning, where unused or redundant connections are eliminated to increase efficiency. The adage "use it or lose it" aptly applies here, as the brain prioritizes pathways that are frequently activated while discarding those that are not. This pruning process is influenced by both genetic programming and environmental input, underscoring the importance of early experiences in shaping brain development. For instance, a child exposed to a rich linguistic environment will strengthen neural circuits related to language processing, while those deprived of such stimuli may struggle to develop robust language skills later in life.

Myelination is another key feature of brain maturation, beginning in infancy and continuing into early adulthood. Myelin is a fatty substance that surrounds axons, the long projections of neurons, enabling electrical signals to travel more quickly and efficiently. The progression of myelination follows a hierarchical pattern, with basic survival functions maturing before higher-order cognitive processes. For example, the brainstem and cerebellum, which regulate vital functions like breathing and coordination, are among the first to myelinate. In contrast, the prefrontal cortex—responsible for complex tasks such as planning, decision-making, and impulse control—is one of the last areas to fully mature.

The transition from childhood to adolescence marks a pivotal phase in brain maturation, characterized by both opportunity and vulnerability. Hormonal changes associated with puberty drive significant remodeling of the brain, particularly in regions related to emotion and reward processing, such as the amygdala and nucleus accumbens. These changes often outpace the development of the prefrontal cortex, which continues to mature well into the mid-20s. This developmental imbalance can explain the heightened emotional reactivity and risk-taking behaviors commonly observed during adolescence. However, adolescence also represents a period of remarkable neuroplasticity, making it an ideal time for learning and skill acquisition.

The maturation of the prefrontal cortex is one of the most defining aspects of brain development during the transition to adulthood. As this region matures, individuals gain greater capacity for executive functions, including working memory, self-regulation, and complex problem-solving. The connections between the prefrontal cortex and other brain regions, such as the limbic system, also become more robust, enabling better integration of emotion and logic. This enhanced connectivity supports the ability to weigh long-term consequences, manage impulses, and navigate social relationships with greater sophistication. By the time the prefrontal cortex reaches full maturity, individuals typically exhibit a level of cognitive and emotional stability that allows them to handle the complexities of adult life.

Even after reaching adulthood, the brain continues to evolve, albeit at a slower pace. Neurogenesis, the generation of new neurons, occurs in specific regions such as the hippocampus, which is associated with memory and learning. This ongoing adaptability reflects the brain's capacity to respond to new experiences, challenges, and environments. Lifelong learning, physical activity, and social engagement have all been shown to promote neuroplasticity, helping to maintain cognitive function and delay the effects of aging. Conversely, factors such as chronic stress, poor nutrition, and lack of stimulation can hinder brain health, underscoring the importance of a supportive lifestyle for optimal brain function.

As individuals age, the brain undergoes subtle changes that can affect cognitive abilities. Processing speed and working memory may decline, but other aspects, such as accumulated knowledge and wisdom, often remain intact or even improve. The brain's ability to compensate for age-related changes through alternative neural pathways demonstrates its remarkable resilience. This adaptability, known as cognitive reserve, highlights the importance of maintaining a healthy brain through proactive measures such as mental stimulation, physical exercise, and social interaction.

The journey of brain maturation is a testament to the complexity and adaptability of the human mind. From the rapid growth of infancy to the refinement of adulthood and the subtle shifts of aging, the brain is in a constant state of change, responding to both internal and external forces. Each stage of development builds upon the previous one, creating a foundation for the unique cognitive and emotional capacities that define the human experience. Understanding this process not only provides insight into the factors that influence brain health but also underscores the potential for growth and resilience at every stage of life.

Chapter 7: Future Perspectives in Neurocognition

Research Technologies

The evolution of research technologies has fundamentally reshaped the way knowledge is pursued, analyzed, and applied across disciplines. Centuries ago, research relied heavily on manual processes, handwritten manuscripts, and painstakingly slow methods of data collection. Today, technological advancements have revolutionized the research landscape, enabling faster, more accurate, and more collaborative efforts. The tools and platforms available have not only streamlined methodologies but have also expanded the horizons of what is possible, pushing the boundaries of discovery to unprecedented levels.

One of the most transformative advancements in research technologies has been the digitization of information. Libraries that once required physical attendance now house vast collections of digitized books, journals, and manuscripts, accessible from virtually anywhere in the world. This shift has democratized access to knowledge, making information available to researchers regardless of geographic or institutional barriers. Digital archives have also preserved historical documents that might otherwise have been lost to time, allowing scholars to study rare materials without the risk of damaging the originals. The integration of search algorithms into these digital platforms has further enhanced their utility, enabling researchers to locate specific information within seconds, a process that would have taken days or weeks in the past.

The advent of specialized software has been another notable milestone. Analytical tools designed for specific fields, from statistical analysis programs to molecular modeling software, have significantly increased the precision and efficiency of research. For instance, in the realm of social sciences, programs like SPSS and R allow researchers to process complex datasets and identify patterns that would be nearly impossible to discern manually. In biological sciences, software capable of simulating molecular interactions facilitates drug discovery and genetic research, reducing the time and resources needed for laboratory experiments. These tools not only enhance productivity but also minimize human error, ensuring more reliable results.

Collaboration has also been transformed by modern research technologies. Cloud-based platforms and real-time communication tools enable researchers from different corners of the globe to work together seamlessly. Collaborative platforms like shared document editors and project management software allow teams to coordinate efforts, discuss findings, and refine methodologies without the constraints of physical meetings. This interconnectedness has fostered interdisciplinary approaches, where experts from diverse fields can converge their knowledge to tackle complex problems, leading to innovations that might not have emerged within the silo of a single discipline.

The role of data in research has grown exponentially, driven by advances in data collection, storage, and analysis technologies. The rise of big data has allowed researchers to work with massive datasets that reveal trends and insights on a scale previously unimaginable. Sensors, satellites, and other data collection devices have enabled the monitoring of phenomena ranging from climate change to urban traffic patterns in real time. The challenge of managing and interpreting these vast amounts of information has given rise to fields like data science and machine learning, which provide researchers with the tools to extract meaningful conclusions from raw data. The ability to visualize data through interactive graphs and simulations has further enhanced comprehension and communication of complex findings.

Ethical considerations have become increasingly important as research technologies advance. The ease of accessing and sharing information has raised concerns about data privacy, intellectual property, and the potential misuse of sensitive information. Researchers must navigate these ethical challenges carefully, ensuring that their work adheres to established guidelines and respects the rights of individuals and communities. Technologies such as encryption and secure data storage systems have been developed to address these concerns, safeguarding the integrity of research and the privacy of its subjects.

Laboratory research has seen its own technological revolution, with automated equipment and robotics transforming experimental procedures. Automated pipetting systems, for example, ensure consistency and precision in chemical and biological experiments, reducing variability and increasing reproducibility. High-throughput screening technologies allow researchers to test thousands of samples simultaneously, accelerating the discovery of new materials, drugs, and biological processes. The integration of robotics in laboratories has not only improved efficiency but has also freed researchers from repetitive tasks, allowing them to focus on analysis and innovation.

Visualization technologies have provided researchers with new ways to interpret and present their findings. Three-dimensional imaging techniques, such as MRI and CT scans, have become indispensable in medical research, allowing detailed views of internal structures without invasive procedures. Virtual reality and augmented reality tools are now being used to simulate environments, model phenomena, and train researchers in complex procedures. These immersive technologies offer unique perspectives that enhance understanding and foster creativity in problem-solving.

The internet has been a game-changer for research dissemination. Open access journals and online repositories have made it easier for researchers to share their findings with the global community. Preprint servers allow for the rapid dissemination of research results, enabling timely discussion and feedback. Social media and academic networking platforms have further amplified the reach of research, connecting scholars with peers and the public alike. This increased visibility has not only facilitated collaboration but has also heightened public engagement with science, bridging the gap between researchers and society.

Emerging technologies continue to hold promise for the future of research. Quantum computing, for instance, has the potential to revolutionize fields such as cryptography, material science, and complex system modeling by performing calculations at speeds far beyond current capabilities. Biotechnology advancements are poised to unlock new frontiers in genetics, agriculture, and medicine, with tools like CRISPR enabling precise gene editing that could address diseases and enhance food security. As these technologies mature, they will undoubtedly open new avenues for discovery and innovation.

The pace of technological progress shows no signs of slowing, and researchers must remain adaptable to harness these advancements effectively. Continuous learning and skill development are essential to keep up with evolving tools and methodologies. By embracing new technologies while maintaining a commitment to ethical practices, researchers can continue to push the boundaries of knowledge and address the pressing challenges of our time. The integration of these technologies into the research process not only enhances efficiency and accuracy but also ensures that the pursuit of knowledge remains at the forefront of human progress.

AI and Neural Networks

The emergence of AI and neural networks represents one of the most profound technological advancements of the modern era. These systems are reshaping industries, redefining our understanding of intelligence, and influencing nearly every aspect of human life. Neural networks, which mimic the structure and function of the human brain, serve as the backbone of many artificial intelligence applications, enabling machines to learn, adapt, and make decisions in ways that were once considered purely within the realm of human cognition.

The concept of neural networks originates from the study of biological neurons, the fundamental units of the human nervous system. Researchers sought to replicate the interconnected nature of these neurons, where each node processes and transmits information to others, forming complex pathways capable of handling vast amounts of data. This inspiration led to the creation of artificial neural networks, which consist of layers of interconnected nodes, or "neurons," designed to process input, identify patterns, and produce outputs.

At the core of these systems is the process of training, where a neural network is exposed to datasets that allow it to recognize patterns and relationships. For instance, in image recognition tasks, a neural network might be trained on thousands of labeled images, enabling it to distinguish between objects such as cars, animals, or buildings. This process involves adjusting the weights of connections between neurons to improve the accuracy of its predictions. These adjustments are made through algorithms like backpropagation, which evaluates errors in the network's output and refines its internal parameters accordingly.

One of the defining characteristics of neural networks is their ability to handle unstructured data, such as text, images, and audio. Traditional computational systems often struggle with such data types, requiring significant preprocessing to make sense of them. Neural networks, however, excel at identifying patterns in complex datasets, making them invaluable in fields like natural language processing, where they enable machines to understand and generate human language. Applications like automated translation, sentiment analysis, and chat systems all owe their capabilities to advances in these networks.

Another pivotal area where neural networks have demonstrated their potential is in the realm of deep learning. Deep learning refers to neural networks with multiple layers, known as deep neural networks. These systems are capable of processing data in increasingly abstract ways, allowing them to tackle complex problems that simpler models cannot. For example, in medical imaging, deep neural networks can analyze scans to detect anomalies such as tumors with remarkable accuracy, assisting healthcare professionals in diagnosing conditions earlier and more effectively.

The versatility of neural networks extends beyond specific applications to a broader spectrum of industries. In finance, they are used for fraud detection, analyzing transaction patterns to identify suspicious activities. In agriculture, they enable precision farming by analyzing weather patterns, soil conditions, and crop health to optimize yields. Autonomous vehicles rely on neural networks to process sensor data, navigate roads, and make split-second decisions. These examples illustrate the transformative potential of AI and neural networks across diverse domains.

Despite their numerous advantages, neural networks are not without limitations. One of their significant challenges lies in the requirement for large amounts of data. For a neural network to perform well, it must be trained on extensive datasets that cover a wide range of scenarios. This dependency can be a hurdle in fields where data is scarce or difficult to obtain. Furthermore, the computational power required to train deep neural networks can be substantial, often necessitating specialized hardware such as graphics processing units (GPUs) or tensor processing units (TPUs).

Another notable challenge is the issue of interpretability. Neural networks often function as "black boxes," meaning their decision-making processes are difficult to understand, even for those who design them. This lack of transparency can be problematic, particularly in high-stakes applications like healthcare or criminal justice, where understanding the rationale behind a decision is critical. Efforts are underway to develop methods for interpreting and explaining neural network outputs, but this remains an area of active research.

Ethical concerns also arise in the deployment of AI and neural networks. Bias in training data can lead to biased outputs, perpetuating or even amplifying societal inequities. For instance, an AI system trained on historical hiring data that reflects gender or racial biases may reproduce those biases in its recommendations. Addressing these ethical challenges requires rigorous scrutiny of datasets, the implementation of fairness measures, and the establishment of guidelines for responsible AI development and use.

The rapid pace of advancements in neural networks is driving innovation, but it also raises questions about the future of work and the societal implications of widespread automation. As machines become capable of performing tasks traditionally reserved for humans, concerns about job displacement and economic inequality are growing. However, the same technologies also have the potential to create new opportunities, enhance productivity, and improve quality of life when integrated thoughtfully and strategically.

As neural networks continue to evolve, their applications are likely to expand into areas we have yet to imagine. The development of generative models, capable of creating new content such as art, music, or even scientific hypotheses, highlights the creative potential of these systems. Furthermore, efforts to combine neural networks with other technologies, such as robotics or quantum computing, could unlock new levels of capability and redefine what machines can achieve.

The intersection of AI and neural networks represents a dynamic and rapidly changing field, marked by both incredible potential and complex challenges. Their impact on society will depend not only on technological advancements but also on the choices we make about how to develop, regulate, and apply these tools. By fostering collaboration between technologists, policymakers, and stakeholders across sectors, we can ensure that the benefits of neural networks are realized in ways that align with broader societal goals and values.

Cognitive Enhancement

The pursuit of cognitive enhancement has intrigued humanity for centuries, driven by the desire to expand the limits of memory, focus, creativity, and overall mental acuity. From ancient herbal remedies to the advanced technologies of today, the concept of improving cognitive function has evolved alongside human understanding of the brain. The modern era has witnessed a convergence of neuroscience, pharmacology, and technology, presenting a variety of approaches that challenge traditional notions of human potential and raise profound ethical questions.

One of the earliest methods of cognitive enhancement involved the use of natural substances to influence mental performance. Ancient civilizations relied on plants like ginseng, gingko biloba, and coca leaves, believing in their abilities to improve focus, energy, or memory. These traditional remedies laid the groundwork for the development of modern nootropics, a broad category of substances designed to enhance cognitive function. Today, nootropics include a range of compounds, from prescription medications like modafinil and methylphenidate to over-the-counter supplements such as omega-3 fatty acids and caffeine. While some of these substances have been rigorously studied and proven effective under specific conditions, others remain controversial, with limited scientific evidence supporting their claims.

Pharmacological advancements have undoubtedly played a significant role in the cognitive enhancement landscape. For individuals with diagnosed conditions like attention deficit hyperactivity disorder (ADHD) or narcolepsy, medications such as Adderall or modafinil can dramatically improve focus and wakefulness. However, the off-label use of these drugs by healthy individuals seeking a mental edge has sparked debates about fairness, dependency, and long-term consequences. The allure of enhanced productivity, particularly in high-pressure environments like academia or corporate settings, has led some to overlook the potential risks, including addiction, side effects, and ethical concerns about accessibility and equity.

Beyond pharmacology, cognitive enhancement has increasingly intersected with technology. Brain stimulation techniques, such as transcranial magnetic stimulation (TMS) and transcranial direct current stimulation (tDCS), have emerged as promising tools for improving cognitive performance. By delivering controlled electrical or magnetic impulses to specific regions of the brain, these methods aim to modulate neural activity in ways that enhance memory, learning, or problem-solving abilities. While still in the experimental stages for many applications, early studies suggest that these techniques could offer non-invasive alternatives to pharmacological interventions, although their efficacy and safety require further investigation.

Neurofeedback is another technological approach gaining traction in the realm of cognitive enhancement. This method involves real-time monitoring of brain activity using electroencephalography (EEG) or other imaging techniques, allowing individuals to learn how to regulate their brainwaves through feedback and practice. Neurofeedback has shown promise in improving attention, reducing stress, and even managing conditions like anxiety or depression. By training the brain to achieve desired states, this approach offers a personalized and self-directed path to cognitive enhancement, free from the side effects associated with drugs or invasive procedures.

Lifestyle factors, too, play a crucial role in shaping cognitive abilities. The impact of diet, exercise, sleep, and mindfulness on brain health is well-documented, underscoring the importance of holistic approaches to enhancement. Regular physical activity has been shown to increase the production of brain-derived neurotrophic factor (BDNF), a protein that supports the growth and maintenance of neural connections. Similarly, a diet rich in antioxidants, healthy fats, and essential nutrients provides the building blocks necessary for optimal brain function. Prioritizing restorative sleep and practicing mindfulness or meditation can further enhance focus, memory, and emotional regulation, offering natural and sustainable ways to boost cognitive performance.

The concept of cognitive enhancement is not limited to individual pursuits; it also has implications for collective progress. Collaborative tools, shared knowledge platforms, and organizational strategies that promote creativity and innovation can amplify the cognitive capabilities of groups, enabling breakthroughs in science, art, and technology. The integration of diverse perspectives and expertise often leads to solutions that transcend the abilities of any single individual, illustrating the potential of cognitive enhancement on a broader societal scale.

Despite the benefits, the pursuit of cognitive enhancement raises ethical and philosophical questions that cannot be ignored. The distinction between therapy and enhancement is often blurred, complicating discussions about what constitutes a "natural" or "acceptable" intervention. For instance, while treating a cognitive deficit is widely regarded as ethical, enhancing an already functional brain to achieve superhuman abilities challenges societal norms and perceptions of fairness. Access to enhancement technologies is another critical concern, as disparities in availability could exacerbate existing inequalities, creating a cognitive divide between those who can afford these advancements and those who cannot.

The long-term consequences of cognitive enhancement remain uncertain, particularly when it comes to altering brain chemistry or neural activity. The brain is a highly complex and interconnected system, and interventions that target specific functions could have unintended ripple effects on other aspects of cognition or behavior. Additionally, the pressure to enhance oneself in competitive environments may lead to overuse or misuse of enhancement tools, raising concerns about mental health, identity, and the definition of success.

The rapid pace of innovation in this field suggests that the future of cognitive enhancement will be shaped by a combination of scientific breakthroughs, technological advancements, and societal values. Emerging fields like neurogenetics, which explores the manipulation of genes to influence cognitive traits, and brain-computer interfaces, which aim to integrate human cognition with digital systems, offer glimpses into possibilities that were once confined to the realm of science fiction. These developments hold immense promise but also demand careful consideration of their ethical, social, and psychological implications.

The journey toward cognitive enhancement is as much about understanding the human brain as it is about pushing its limits. By balancing ambition with responsibility, innovation with ethics, and individual goals with collective well-being, society can navigate this complex landscape in ways that benefit humanity as a whole. Whether through natural means, pharmacological aids, or technological interventions, the pursuit of enhanced cognition reflects a fundamental aspect of human nature: the desire to grow, adapt, and achieve beyond perceived boundaries.

Therapeutic Applications

The field of therapeutic applications has undergone a transformative evolution, driven by advancements in science, medicine, and technology. These innovations have redefined how diseases are treated, how recovery is approached, and how individuals can achieve an improved quality of life. Therapeutic practices now span a vast array of disciplines, from traditional methods rooted in centuries of observation to cutting-edge techniques that leverage molecular biology and precision medicine. The convergence of various approaches has created a dynamic landscape that continues to expand the possibilities for healing and restoration.

One of the most impactful developments in therapeutic applications has been the rise of personalized medicine. This approach tailors treatments to the unique genetic makeup and characteristics of each patient, moving away from the one-size-fits-all model that has dominated much of medical history. Advances in genomics have made it possible to identify specific mutations or biomarkers that influence the progression of diseases such as cancer. By understanding these molecular drivers, therapies can be designed to target them directly, minimizing side effects and maximizing efficacy. For example, targeted therapies like tyrosine kinase inhibitors have revolutionized the treatment of certain cancers by blocking the proteins that fuel tumor growth, offering patients a more effective and less invasive alternative to traditional chemotherapy.

Immunotherapy has also emerged as a cornerstone of modern therapeutic applications. By harnessing the body's natural defense system, immunotherapy has demonstrated remarkable success in treating conditions that were once considered untreatable. Techniques such as checkpoint inhibitors, which remove the "brakes" on immune cells, and CAR-T cell therapy, which engineers a patient's own immune cells to recognize and destroy cancer cells, have redefined the possibilities of treatment. These breakthroughs not only represent a paradigm shift in oncology but have also inspired research into their use for autoimmune disorders, infectious diseases, and even neurological conditions.

Regenerative medicine has further expanded the scope of therapeutic applications, focusing on repairing or replacing damaged tissues and organs. Stem cell therapy, a key component of this field, has shown promise in treating a wide range of conditions, from spinal cord injuries to degenerative diseases like Parkinson's. By using stem cells' ability to differentiate into various cell types, researchers are exploring ways to restore function in damaged areas of the body. Similarly, advances in tissue engineering and 3D bioprinting have opened up the possibility of creating lab-grown organs, potentially eliminating the need for donor transplants and reducing the risk of rejection.

In the realm of chronic diseases, therapeutic applications have increasingly emphasized a holistic approach that integrates physical, mental, and emotional well-being. Conditions such as diabetes, cardiovascular disease, and chronic pain often require long-term management rather than a one-time cure. Lifestyle interventions, including dietary modifications, exercise programs, and stress management techniques, play a crucial role in these therapeutic strategies. Cognitive-behavioral therapy (CBT) and other psychological interventions have also become essential components of treatment plans, particularly for conditions like depression and anxiety that often accompany chronic illnesses.

Technological advancements have introduced new dimensions to therapeutic applications, particularly through wearable devices and remote monitoring. These tools enable continuous tracking of vital signs, medication adherence, and other health metrics, allowing for timely interventions and more personalized care. For instance, wearable cardiac monitors can detect irregular heart rhythms and alert healthcare providers in real time, potentially preventing life-threatening events. Similarly, remote monitoring systems for patients with diabetes or hypertension provide actionable insights that help individuals and their care teams make informed decisions.

The integration of rehabilitation therapies into treatment plans has also seen significant progress. Physical therapy, occupational therapy, and speech therapy are no longer limited to traditional methods but now incorporate advanced tools and techniques to enhance recovery. Robotic exoskeletons, for example, are being used to assist individuals with mobility impairments, enabling them to regain movement and independence. Virtual reality (VR) has found applications in rehabilitation as well, providing immersive environments that motivate patients to engage in therapeutic exercises while tracking their progress in real time.

The treatment of mental health disorders has greatly benefited from advancements in therapeutic applications, with a growing emphasis on destigmatizing these conditions and improving access to care. Psychopharmacology has developed medications that target specific neurotransmitter imbalances, offering relief for conditions such as depression, bipolar disorder, and schizophrenia. At the same time, psychotherapeutic approaches have evolved to include evidence-based interventions like mindfulness-based stress reduction (MBSR) and dialectical behavior therapy (DBT). Emerging therapies, such as psychedelic-assisted therapy, are also gaining attention for their potential to address treatment-resistant conditions, though they remain an area of active research and debate.

Therapeutic applications have extended beyond the confines of hospitals and clinics, embracing community-based and home-based care models. These approaches recognize the importance of social support and environmental factors in healing. Palliative care, for instance, focuses on improving the quality of life for individuals with serious illnesses by addressing their physical, emotional, and spiritual needs. Home-based care programs enable patients to receive treatments in familiar surroundings, reducing stress and improving compliance with therapeutic regimens. This shift toward patient-centered care reflects a broader understanding of health and well-being as interconnected and multifaceted.

While the advancements in therapeutic applications are undoubtedly remarkable, they also raise important ethical and practical considerations. The high cost of cutting-edge treatments, for example, often limits their accessibility, creating disparities in healthcare outcomes. Ensuring equitable access to these innovations remains a critical challenge for policymakers, healthcare providers, and researchers alike. Additionally, the rapid pace of development sometimes outstrips the regulatory frameworks designed to ensure safety and efficacy, necessitating robust systems for evaluating and monitoring new therapies.

The interplay between innovation and tradition is another intriguing aspect of therapeutic applications. While modern medicine continues to push the boundaries of what is possible, there is a growing recognition of the value of traditional healing practices and their potential to complement contemporary treatments. Integrative medicine seeks to bridge this gap, combining evidence-based practices from both conventional and alternative approaches to provide comprehensive care.

The journey toward advancing therapeutic applications is one of constant exploration, driven by the needs and aspirations of patients and the dedication of those who seek to improve their lives. Each breakthrough represents a step forward in understanding the complexities of the human body and mind, as well as the intricate interplay between biology, environment, and society. As the field continues to evolve, it holds the promise of not only treating diseases but also enhancing resilience, extending longevity, and fostering a deeper connection to the essence of health and healing.

Ethics in Neuroscience

Ethics in neuroscience is a field that navigates the intricate intersection of scientific discovery, human values, and societal impact. As research delves deeper into understanding the brain and its complexities, the ethical considerations surrounding these advancements grow increasingly significant. The study of neuroscience holds immense promise for treating neurological disorders, enhancing cognitive abilities, and unraveling the mysteries of human behavior. Yet, these possibilities bring with them profound questions about consent, privacy, identity, and the boundaries of intervention.

One of the most pressing ethical concerns in neuroscience revolves around the concept of informed consent. Neuroscientific research often involves vulnerable populations, including individuals with cognitive impairments, psychiatric conditions, or brain injuries. Ensuring that participants fully understand the nature of the research, its potential risks, and its implications is a cornerstone of ethical practice. However, cognitive limitations or altered mental states can complicate this process, raising questions about whether true consent can be obtained. Researchers must carefully balance the need for scientific progress with the imperative to protect the autonomy and dignity of participants.

The issue of privacy becomes particularly significant as neuroscience advances in brain imaging and data collection. Technologies such as functional magnetic resonance imaging (fMRI) and electroencephalography (EEG) allow researchers to observe brain activity in unprecedented detail, revealing information about thoughts, emotions, and decision-making processes. While these insights have transformative potential, they also raise concerns about the potential misuse of such data. For example, could employers, insurers, or governments use brain data to make decisions about hiring, coverage, or security clearance? Safeguarding the confidentiality of neural information and establishing clear guidelines for its use are critical to preventing abuses and ensuring that individuals retain control over their most intimate mental processes.

The concept of neuroenhancement introduces another layer of ethical complexity. Technologies and interventions designed to improve cognitive function, memory, or mood blur the line between therapy and enhancement. While treating neurological disorders is widely accepted as ethical, enhancing the abilities of healthy individuals raises questions about fairness, societal pressure, and the potential for unintended consequences. For instance, if neuroenhancement becomes widely available, will it create a divide between those who can afford these interventions and those who cannot? Will individuals feel compelled to enhance themselves to remain competitive in academic, professional, or social settings? These concerns highlight the need for thoughtful regulation and societal dialogue about the implications of neuroenhancement.

Neuroscience also challenges traditional notions of personal identity and free will. As researchers uncover the neural mechanisms underlying behavior and decision-making, they raise questions about the extent to which individuals are truly autonomous. If certain behaviors can be traced to specific patterns of brain activity or genetic predispositions, does this diminish personal responsibility? These questions are particularly relevant in the context of criminal justice, where neuroscientific evidence is increasingly being used to assess culpability and inform sentencing. While such evidence can provide valuable insights, it also raises concerns about determinism and the potential for bias in interpreting complex neural data.

The use of neuroscience in marketing and persuasion, often referred to as neuromarketing, presents another set of ethical challenges. By studying how the brain responds to stimuli, marketers can design strategies that influence consumer behavior more effectively. While this has clear benefits for businesses, it also raises concerns about exploitation and manipulation. If companies can tap into subconscious processes to drive purchasing decisions, does this undermine consumer autonomy? Additionally, the use of neuroscientific techniques to influence political campaigns or public opinion poses risks to democratic processes, highlighting the need for transparency and ethical guidelines in the application of such methods.

Animal research remains a contentious issue within neuroscience, as many studies rely on animal models to investigate brain function and develop treatments. While these studies have led to significant medical advances, they also raise questions about the ethical treatment of animals and the justification for their use. Researchers are tasked with minimizing harm and ensuring that experiments are scientifically necessary, but the debate over the moral status of animals and the balance between human benefit and animal welfare persists. The development of alternative methods, such as organoids or computer simulations, offers potential pathways to reduce

reliance on animal research, though these approaches
are still in their infancy.

Emerging technologies, such as brain-computer
interfaces (BCIs) and neural implants, present both
incredible opportunities and profound ethical
dilemmas. These devices have the potential to restore
mobility to individuals with paralysis, improve
communication for those with severe disabilities, and
even augment cognitive function. However, they also
raise questions about consent, long-term safety, and
the potential for misuse. For example, who owns the
data generated by neural devices, and how can it be
protected? What happens if a neural implant
malfunctions or is hacked? These questions
underscore the importance of proactive ethical
frameworks to guide the development and
deployment of such technologies.

The global nature of neuroscience research adds
another layer of complexity to ethical considerations.
Different cultures and societies may have varying
perspectives on issues such as neuroenhancement,
privacy, and the use of brain data. International
collaboration is essential for advancing neuroscience,
but it also requires navigating these cultural
differences and ensuring that ethical standards are
upheld across borders. Establishing universal
principles while respecting cultural diversity is a
delicate but necessary task for the global neuroscience
community.

Education and public engagement play a crucial role
in addressing the ethical challenges of neuroscience.
As the field progresses, it is essential to foster an

informed and inclusive dialogue about the implications of neuroscientific discoveries. Public understanding of neuroscience can empower individuals to make informed decisions about their own health and contribute to shaping policies that reflect societal values. Additionally, involving diverse perspectives in ethical discussions can help ensure that the benefits of neuroscience are distributed equitably and that potential harms are minimized.

The rapid pace of discovery in neuroscience holds immense promise for improving lives and expanding our understanding of what it means to be human. However, it also demands a thoughtful and proactive approach to ethics, one that anticipates challenges, protects individual rights, and promotes the responsible use of knowledge. As the boundaries of neuroscience continue to expand, the integration of ethical considerations into every stage of research and application will be essential to ensuring that these advancements serve the greater good. The questions raised by neuroscience are not merely scientific—they are deeply human, reflecting our values, our aspirations, and our responsibility to one another.